SpringerBriefs in Medical Earth Sciences

More information about this series at http://www.springer.com/series/13202

Kirtikumar Randive · Pratik Godbole

Medical Geology
for Beginners

 Springer

Kirtikumar Randive
Department of Geology
RTM Nagpur University
Nagpur, Maharashtra, India

Pratik Godbole
Department of Geology
RTM Nagpur University
Nagpur, Maharashtra, India

ISSN 2523-3610   ISSN 2523-3629 (electronic)
SpringerBriefs in Medical Earth Sciences
ISBN 978-3-031-82764-8   ISBN 978-3-031-82765-5 (eBook)
https://doi.org/10.1007/978-3-031-82765-5

This Springer imprint is published by the registered company Springer Nature Switzerland AG
The registered company address is: Gewerbestrasse 11, 6330 Cham, Switzerland

If disposing of this product, please recycle the paper.

You saw a dream for me…. and made sure that I realize it. Through the rough seas of life, you kept me afloat…. thousands of words can never express how I felt when you were around! …. Now the journey is over…

—Kirtikumar Randive lost his mother on 7th October 2024 while he was writing the manuscript of this book.

Foreword

Medical Geology has been defined as the impacts of geologic materials and geologic processes on animal and human health. As a formal discipline medical geology can be traced back to the pioneering efforts of Olle Selinus and his colleagues and their award winning book *Essentials of Medical Geology*. Since then at least 40 books on medical geology, including two by the author of this book, have appeared in at least six languages. There have been medical geology short courses, conferences, university credit courses, webinars, conference technical sessions, numerous published reports addressing a wide range of medical geology issues, as well as several scientific organizations dedicated to this discipline. However, there has been a lack of efforts to bring this important issue to the attention of the next generation of scientists. The generation that we hope will help to minimize the debilitating medical geology issues that impact many millions of people in every corner of the world.

Medical Geology for Beginners successfully fills this gap by describing medical geology is terms that non-professionals can easily understand and by profusely illustrating the book with attractive, informative illustrations. In easy-to-understand terms, the book describes a wide range of medical geology issues including both the beneficial and deleterious impacts of geologic materials. One of the most important contributions of the book is that it clearly explains the important roles for medical geologists in addressing these environmental health problems that afflict countless people around the world.

Medical Geology for Beginners provides an invaluable service to the science of medical geology and to anyone interested in this subject matter and especially to the young readers who may be inspired to pursue a career in medical geology.

Dr. Robert B. Finkelman
US Geological Survey, Retired,
Research Professor
University of Texas at Dallas
Richardson, USA

Preface

It is a great pleasure that we are coming up with a fully dedicated monograph on Medical Geology; this time for the students and enthusiasts who find this new domain of science worth exploring. The title itself suggests that the book is intended for those who are new to this discipline and the target audience are the students who are studying Medical Geology for their undergraduate and post-graduate courses. Our first book *Elements of Geochemisty, Geochemical Exploration and Medical Geology* was published more than 12 years ago, and then a dedicated volume entitled *Medical Geology in Mining: Health Hazards due to Metal Toxicity* followed in 2022. These books were dedicated to the researchers and experts in this field. However, the great majority of readers are the students who find medical geology as an exciting subject of their academic interest. Therefore, it was felt that there should be a book that would cater to the young minds and quench the thirst for knowledge on this subject in a manner in which they could relate and understand. We discussed this idea with Alexis Vizcaino who found it very interesting and connected us to Qiao Shu who gladly took up this project and made sure that it was completed in time. She also suggested Springer brief series since the intended readers are mostly beginners in this subject. Finally, we agreed on the title and the project was completed by the end of October 2024.

The present book is divided into five chapters in which, first Chap. 1 introduces the fundamental concepts of the medical geology, its scope, historical background, and prospects. Chapter 2 emphasizes on the trace elements (essential and probably-essential) and their requirement for human body. Chapter 3 delves into toxicity caused due to certain trace elements. Chapter 4 elaborately discusses the diseases caused due to geological sources, and Chap. 5 discusses therapeutic potential of geo-materials. Towards the end of each chapter, several references are given to provide a detailed bibliography of available literature for the students and researchers. There are a number of beautiful illustrations and useful tables. We hope that the captivating sketches and simple narration will make this book an interesting reading for all the readers of this book.

Nagpur, India   Kirtikumar Randive
October 2024   Pratik Godbole

Acknowledgements

The idea of this book was first discussed with Alexis Vizcaino who is always excited and positive about new book projects and responded very positively. A good friend in him is always acknowledged and thanked for his support and guidance. Associate editor of Springer Dr. Qiao Shu reviewed this proposal and made several useful suggestions for which we are grateful to her. We also thank Mr. Ravi Vengadchalam for his prompt reminders and support during various stages of publication of this book. We are very much thankful to Prof. Robert Finkelman for writing a forward for this book. We are thankful for exciting coffee table discussions with our enthusiastic team members, Dr. Sanjeevani Jawadand, Dr. Sneha Dandekar, Tejashree Raut, Samiksha Bawangade, Kaustubh Deshpande, Pratiksha Wankhede, Krutika Jangale, Pooja Nandi, Gaurav Pawar, Kiran Surve, Martina Malluri, Mayank Joshi, Radhika Rajdhar, Ritik Palaspagar, Tejal Nirwan, Parul Dhakate, Saili Dhok, Pranay Meshram, Tanushree Dupare, Shruti Meshram, Parag Janbandhu, Vaishnavi Tikde, Atul Selokar, Gopal Daware, Kishor Deshmukh, Avantika Singh, Ajay Padvi, Akshay Naitam, Chitra Pipaldhare, Tanushree Nikhure, Shreya Meshram, and Pooja Ilamkar. We also thank our colleagues and friends Dr. M. L. Dora, Dr. Rajkumar Meshram, Dr. Tushar Meshram, Dr. J. Vijaya Kumar, Dr. Manoj Sahu, Dr. Shubhangi Lanjewar, Dr. Pankaj Kumar, Diksha Khandelwal, Rajeev Kumar, Chaitya Aswal, Dr. Apurva Barve, Dr. Yogita Mahajan, Dr. Mahesh Korakoppa. We are also thankful to Alecos Demetriades, Carlos Monteiro, and Matteo Serra for inviting us for the 3rd International Student Conference on Medical Geology and Environmental Health initiating fruitful discussions and ideas. We are very thankful to Gargee Deshpande for working as an illustrator and designer on this project. We also thank Rajesh Landge, Yukti Waghale and Sarang Pradhan for their help and support. Finally, we are indebted to our family members Chetana Randive, Raghav Randive, Rashmi Godbole, Pravin Godbole, Harshada Godbole, Pranjal Godbole, Nazia Godbole, Aahan Godbole, Sana Pandit, Samir Pandit, Sagar Pandit, Sahil Pandit, Kapil Kher, Dr. Sanjeev Gokhale, Rajendra Adhikari and Makarand Bhagwat.

Contents

Chapter 1
Fundamentals of Medical Geology

Medical geology explores how geological materials affect human health. This chapter introduces medical geology and discusses how geological factors and climate can trigger diseases. The Geo-medical cycle shows how elements from rocks and human activities enter our bodies and impact health through interactions between the rock, water, and biological systems. This chapter also explains how weathered rocks release important nutrients as well as harmful trace element which affect human health. It further highlights the role of micronutrients in plant growth and their availability in the soil. Finally, the chapter discussed Geo-pharmacy, showcasing how geological materials are used as medicines.

1.1 What Is Medical Geology?

Medical geology is the study of geological factors affecting the health of humans and other animals. It was first defined as "*a science that deals with the relationship between natural geological factors and health in humans and animals, and understanding how everyday environmental factors impact health in different geographic regions*" (Randive 2013). This field combines knowledge from different sciences, fostering collaborations between various disciplines, especially geology and medical science, to address the health-related challenges (Selinus 2019). Medical geology connects geology to health, environment and the anthropological activities in order to understand the mechanism of transport of the chemical elements that are released in the environment from the geologic sources. In 2003, the International Medical Geology Association (IMGA) defined medical geology as "*the science that focuses on the relationship between geological factors and health problems in living organisms.*" However, this definition was considered too narrow because it did not encompass key environmental factors such as air, water, and soil (Hasan 2021). To make it more comprehensive, Hasan (2019) proposed a revised definition, which also considers global climate change and its effect on the public health and ecosystems. This updated definition includes both human-made (anthropogenic) and natural geological factors, and their impact on human and ecological health (Hasan 2019).

1.2   The Historical Connect of Geology and Health

Throughout history, ancient civilizations such as India, Greece, China, and Rome recognized a connection between geology and health. The writings of Hippocrates and ancient Chinese medical texts reflect this understanding, showing awareness of how environmental factors like heavy metals and dietary imbalances could affect health. Historical cases, such as lead poisoning and iodine deficiency, provide evidence of these connections. Archaeological findings, including the remains of ancient civilizations, further demonstrate the link between environmental conditions and health. These findings shed light on past epidemics and nutritional deficiencies. The transition from hunter-gatherer societies to farming-based communities also influenced health, with dietary changes leading to issues such as iron deficiency (anaemia) (Selinus et al. 2019). Minerals have played a significant role in medicine across many cultures. In ancient Chinese and Indian civilizations dating back to around 3000 BC, minerals like cinnabar, galena, and realgar were used for their healing properties. Elements such as gold, silver, iron, and zinc were also used, often in combination with herbs. Similarly, civilizations like the Assyrians, Babylonians, and Egyptians used natural substances such as alum, bitumen, copper, and potassium nitrate in their medicinal practices.

Records show that the Greeks around 400 BC and the Mayans around 800 AD also used minerals in their medicines. During the Islamic Golden Age, scholars such as Rhazes, Abulcasis, and Avicenna made significant advancements in medicine. Rhazes, in his "Comprehensive Book of Medicine," described use of alum, saltpetre, gold, and mercury in treatments (Hasan 2021). Abulcasis, in his "Kitab al-Tasrif," detailed medical techniques involving plants, minerals, and animal products (Sadra 2013). Avicenna's works, especially the "Canon of Medicine," were influential in Europe and emphasized the relationship between the environmental factors and health. He described 760 different drugs derived from plants, rocks and minerals (Gitano 2024) and discussed the spread of diseases through air, water, and soil (Bryne 2012; Hasan et al. 2013).

1.3   Re-emergence of Medical Geology

Medical geology is not a new field but rather a revival of an ancient discipline. The connection between geological materials like rocks and minerals and human health has been recognized for centuries. Ancient civilizations have documented both the healing properties and potential dangers of various rocks and minerals. For example,

Chinese texts from over 2,000 years ago described medicinal uses of around 46 different minerals (Finkelman et al. 2005). Originally, the term "geomedicine" was given for this emerging field. Researchers like J. Lag from the Geological Survey of Norway had used it to describe the link between geology and health (Lag 1990). However, this terminology was not widely accepted because it suggested a subspecialty within medicine, similar to fields like family medicine or nuclear medicine. Instead, the term "medical geology," which was first mentioned in 1834 by an anonymous British physician, was officially adopted in 1997 in a meeting of the Medical Geology Working Group at the 4th International Symposium on Environmental Geochemistry in Vail, Colorado, USA. Professor Olle Selinus, a member of the Commission on Geological Sciences for Environmental Planning (COGEOENVIRONMENT) of the International Union of Geosciences (IUGS), initiated a proposal for forming a working group on medical geology in 1996. The IUGS approved this proposal and appointed Selinus as the chair of the group. During the Vail meeting, which brought together geoscientists, public health professionals, and medical scientists, the term "medical geology" was unanimously chosen. It was decided that "geomedicine" did not fit well with the goals of the discipline or with the medical and public health communities (Selinus 2019, 2020) (Fig. 1.1).

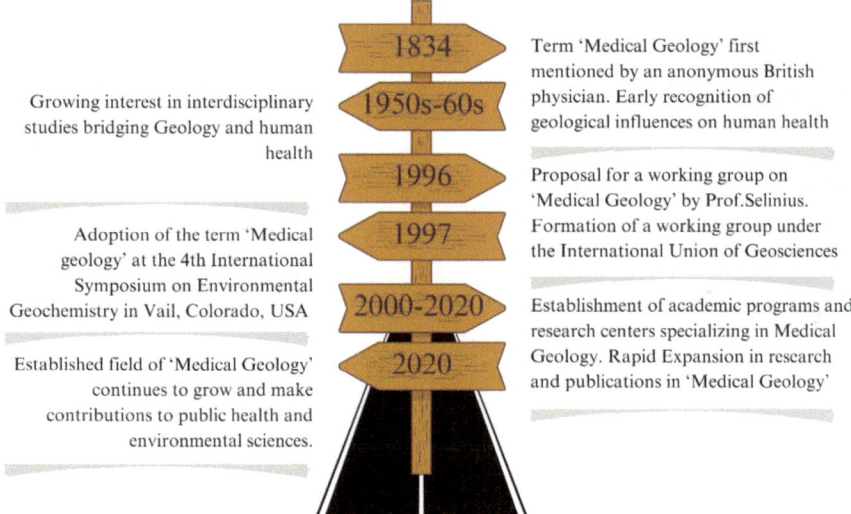

Fig. 1.1 Showing the origin and conceptualisation of Medical Geology through time (Modified after Hasan (2021))

1.4   Current Landscape and Scope of Medical Geology

Medical Geology brings together geoscientists, medical researchers, and public health experts to address health issues resulting from continuous exposure to toxic elements sequestered in the geological reservoirs such as rocks and minerals, and released through the atmospheric dust and water during mining activities. Similarly, the geological events like volcanic eruptions and earthquakes also release these harmful elements (Selinus et al. 2005). Medical geology seeks to understand how people are exposed to these hazards and develop strategies to reduce or prevent this exposure (Sunitha and Reddy 2012). This interdisciplinary field offer numerous opportunities for protecting human as well as ecological health. The scope of medical geology spans from microscopic to large-scale impacts. For instance, it explores how geological factors influence the diseases like cardiovascular disorder and Alzheimer's disease, examines the antibiotic properties of clay minerals, and monitors diseases related to climate change. Medical geologists play a crucial role in the global health management, contributing to multidisciplinary teams involved in public education, citizen science, policy-making, and dissemination of critical information (Hasan 2021). The field also focuses on identifying and characterizing both natural and manmade sources of harmful materials in the environment. This includes predicting the movement and transformation of toxic, infectious, and disease-causing agents over time and space. Currently, medical geology continues to tackle these challenges by integrating geoscientific and health-related knowledge to effectively mitigate the environmental health risks (Fig. 1.2).

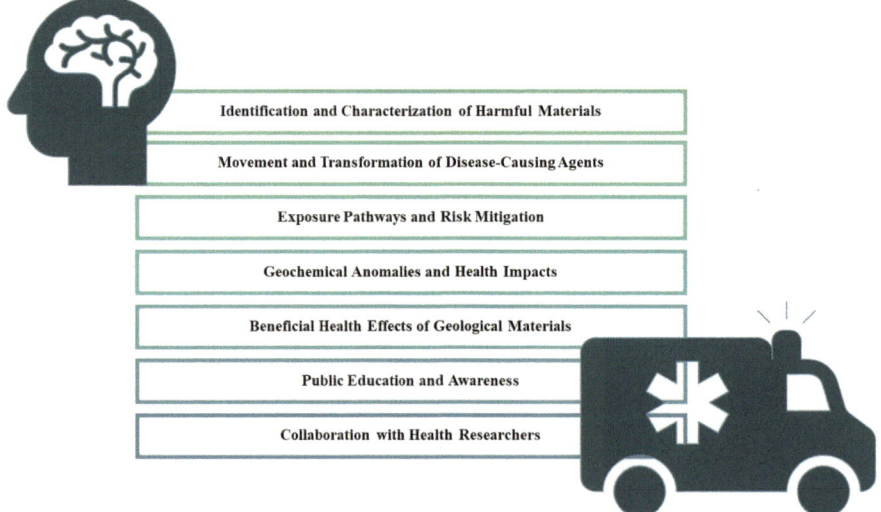

Identification and Characterization of Harmful Materials

Movement and Transformation of Disease-Causing Agents

Exposure Pathways and Risk Mitigation

Geochemical Anomalies and Health Impacts

Beneficial Health Effects of Geological Materials

Public Education and Awareness

Collaboration with Health Researchers

Fig. 1.2 Showing scope of medical geology in current landscape (Modified after Sunitha and Reddy (2012), Hasan (2021), Selinus et al. (2005))

1.5   Role of a Medical Geologist

A medical geologist studies the connection between geology and human health. To do this, he needs a basic understanding of human anatomy and physiology. This helps him to understand how certain elements from the earth (geogenic sources) can either harm or benefit the human body when transported and stored inside (Randive 2013). Furthermore, a medical geologist studies how the geological factors affect the health of humans, animals, and plants directly or indirectly. He also explores how natural resources can be used in modern medicine and diagnostic tools (Geological Survey Ireland 2024). The goal of a medical geologist is to understand sources of toxic elements in the environment and the mechanism of their transport through the biomass, soil and water before entering a human body. This process is called the 'geomedical cycle'. Understanding how certain elements travel from rocks into the human body is a key part of his work. He also studies diseases caused by natural disasters and he works on finding ways to reduce the risks of such diseases. Another important area for a medical geologist is "Geopharmacy", which involves use geological materials to benefit human health (Randive 2013). The future of medical geology is bright. This field is now contributing substantially to the new-age health challenges by way of identifying new geogenic sources of hazards through research e.g. Randive et al. (2022). Similarly, substantial contribution is being made in the field of climate change and human health. For achieving these goals, a medical geologist should create global collaborations, and participate in framing policies. Medical geologist should use advanced technology to study spread of geological hazards and their health effects, create strategies to address climate-related health risks, and set up monitoring mechanism to track environmental health risks (Selinus 2007). He will also play an important role in public health education, preparation for alarm and subsequent response to disasters, and promoting the use of earth materials in modern medicine. By doing this, a medical geologist can greatly improve public health, strengthen disaster response, and support sustainable healthcare practices (Fig. 1.3).

1.6   The Geo-Medical Cycle

The Geo-medical cycle explains how chemical elements released from geogenic and anthropogenic sources enter the human body and impact health (Randive 2013). This cycle shows how different systems—the rock cycle, hydrological cycle, biological cycle, and biochemical cycle—interact with each other to transport contaminants and pathogens. The cycle starts with the rock cycle (Ahmed et al. 2017). Over time, rocks break down into smaller particles through weathering, forming soil. Soil is essential for growing plants and is the foundation of the food chain (Goldhaber and Banwart 2015). The hydrological cycle connects with the rock cycle by distributing water across Earth. Water helps to carry the weathered material into the soil and also transports contaminants (Sasan et al. 2020). These contaminants enter the water

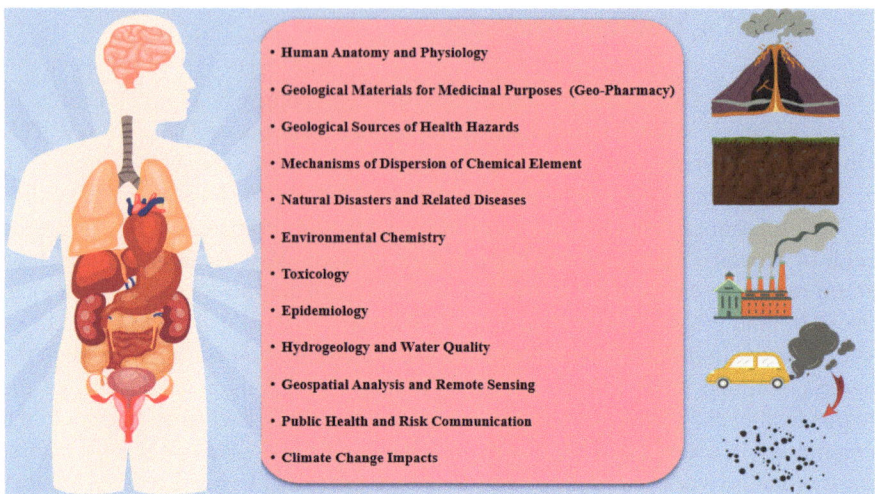

Fig. 1.3 Diagram illustrating the diverse fields of knowledge and competencies a medical geologist must acquire (Modified after Randive (2013))

from various sources, including industrial waste, agricultural runoff, and air pollution. As water moves through rivers, lakes, and oceans, it distributes contaminants (Godbole et al. 2024; Bashir et al. 2020; Nzengung and Gugolz 2022). Plants absorb water and nutrients from the soil. Unfortunately, they can also take in contaminants, which then enter the food chain (Hou and Connor 2020). Animals eat these plants, and contaminants build up in their bodies. This process continues up the food chain until humans are affected when we eat contaminated plants and animals (Chojnacka and Mikulewicz 2024). The biochemical cycle deals with the movement of chemicals between living organisms and the environment (Huang et al. 2024). Harmful substances like heavy metals, pesticides, and pharmaceutical wastes can upset this balance leading to ecological problems and health risks (Asiminicese et al. 2024). Human waste—radioactive, solid, biomedical, and pharmaceutical—also adds to this contamination (Randive et al. 2023; Godbole et al. 2024). Pathogens, like bacteria and viruses, are also part of this cycle. They come from both natural processes and human activities. Soil naturally contains many pathogens, but farming (agriculture), municipal waste, and biomedical contamination can also introduce them. Pathogens can spread through water and soil, entering the food chain and causing diseases in humans. Human activities such as mining, agriculture, and transportation worsen contamination. Mining releases harmful substances into the environment. Agriculture uses pesticides and fertilizers for crop growth, which seeps into the soil and water. Their transportation creates air pollutants (Godbole et al. 2024). These human activities contribute to anthropogenic contamination. These harmful chemicals can enter the human body by breathing contaminated air, drinking contaminated water or eating contaminated food (National Research Council 2000a, b; Kumar et al. 2024). Workers in industries like mining and transportation face more risks (Tumane

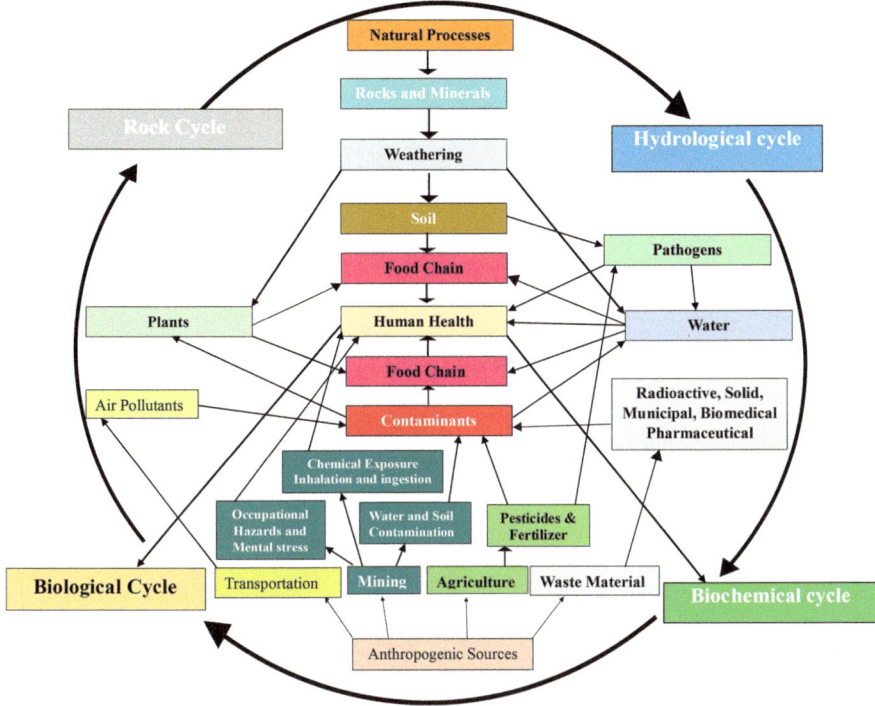

Fig. 1.4 The Geomedical cycle showing transport of elements from rocks and anthropogenic sources to human body (Modified after Randive (2013))

et al. 2022). In summary, the Geo-medical cycle shows the path and mechanism of transport of contaminants and pathogens through soil, water, plants, and animals, eventually affecting human health (Wang et al. 2024) (Fig. 1.4).

1.7 Mobility of Chemical Elements in the Geological Environment

When rocks break down, they release essential nutrients like calcium, magnesium, and potassium, as well as trace metals such as arsenic and lead. Beneficial elements like calcium and magnesium can promote plant growth and contribute to better health, such as stronger bones (Klaes et al. 2022). However, harmful trace elements can also pose health risks; for instance, exposure to lead can lead to serious health issues. In medical geology, the therapeutic uses of geological materials are highlighted through practices like mud therapy, which uses clay minerals to treat skin conditions, and balneotherapy, which involves bathing in mineral-rich hot springs. Additionally, the minerals from weathered rocks influence the quality of natural waters, making

Table 1.1 Relative stability of common igneous rock minerals and associated trace elements (after Randive 2013).

Stability	Mineral	Major constituents	Trace constituents
Easily weathered	Olivine	Mg, Fe, Si	Ni, Co, Mn, Li, Zn, Cu, Mo
	Hornblende	Mg, Mg, Fe, Ca, Al, Si	Ni, Co, Mn, Sc, Li, V, Zn, Cu, Ga
	Augite	Ca, Mg, Al, Si,	Ni, Co, Mn, Sc, Li, V, Zn, Pb, Cu, Ga.
	Biotite	K, Mg, Fe, Al, Si	Rb, Ba, Ni, co, Sc, Li, Mn, V, Ga
	Apatite	Ca, P, F	Rare earths, Pb, Sr
	Anorthite	Ca, Al, Si	Sr, Cu, Ga, Mn
	Oligoclase	Na, Ca, Al, Si	Cu, Ga
Moderately stable	Albite	Na, Al, Si	Cu, Ga
	Garnet	Ca, Mg, Fe, Al, Si	Mn, Cr, Ga
	Orthoclase	K, Al, Si	Rb, Ba, Sr, Cu, Ga
	Muscovite	K, Al, Si	F, Rb, Ba, Sr, Ga, V
	Titanite	Ca, Ti, Si	Rare earths, V,Sn
	Ilmenite	Fe, Ti	Co, Ni, Cr, V
	Magnetite	Fe	Zn, Co, Ni, Cr, V
	Tourmaline	Ca, Mg, Fe, B, Al, Si	Li, F, Ga
	Zircon	Zr, si	Hf
Very stable	Quartz	Si	

them beneficial for drinking and improving health (Carlos 2021). Understanding how weathering affects the mobility of these elements is essential for assessing their positive and negative impacts on human health and the environment (Tables 1.1 and 1.2).

1.7.1 Plants and Micronutrients

Micronutrients are trace elements that are essential for plant growth. These include boron (B), cobalt (Co), chlorine (Cl), copper (Cu), iron (Fe), manganese (Mn), molybdenum (Mo), nickel (Ni), sodium (Na), and zinc (Zn). While these elements help plants grow, some trace elements can be harmful, such as lead (Pb), mercury (Hg), arsenic (As), selenium (Se), cadmium (Cd), and chromium (Cr). The toxicity of these elements does not always depend on how much is present in the plant. Instead, it

Table 1.2 Deposition of sedimentary rocks with their associated trace elements (Randive 2013)

	Major constituents	Types of products	Main Rock Types	Associated trace elements
Process of sedimentation	Si	Resistates	Sandstones	Zr, Ti, Sn, Rare earths, Th, Au, Pt. etc.
			Shales and bituminous shales	V, U, As, Sb, Mo, V, U, As, Sb, Mo, Cu, Ni, Co, Cd, Ag, Au, Pt, B,Se
	Al Si K	Hydrolyzates		
			Bauxites	Be, Ga, Nb, Ti
			Iron ores	V, P, As, Sb, Mo, Se
	Fe Mn	Oxidates	Manganese ore	Li, K, Ba, B, Ti, W, Co, Ni, Cu, Zn, Pa
	Ca, Mg Fe K	Carbonates	Limestones, dolomites	Ba, Sr, Pb, Mn
	Na, Ca, Mg	Evaporates	Salt Deposits	B, I
	Na Mg	Sea		B, Al, I, Br, F, Rb, Li

relies on several processes that determine how available and soluble these elements are in the soil. These processes include (Share 2024a, b):

1. *Desorption or Dissolution*: Elements are released from solid materials into the soil solution.
2. *Chemical Speciation*: Different forms of an element can affect its reactivity and movement.
3. *Diffusion or Mass Flow*: Elements move through soil solutions towards plant roots.
4. *Sorption or Precipitation*: Elements bind to soil particles or form insoluble compounds.
5. *Absorption by Roots*: Plant roots take up elements from the soil.
6. *Translocation from Roots to Tops*: Absorbed elements move from the roots to other parts of the plant, like leaves (Figs. 1.5 and 1.6).

1.8   An Introduction to Geo-Pharmacy

Rocks and minerals have always been a backbone of human civilization. Their uses are manyfold, such as making of tools, artefacts, ornaments, industrial appliances as well as medicines. In India, ancient texts about Ayurveda, written by scholars like Charak, Sushrut, and Dhanvantari, mentioned the use of different geomaterials such

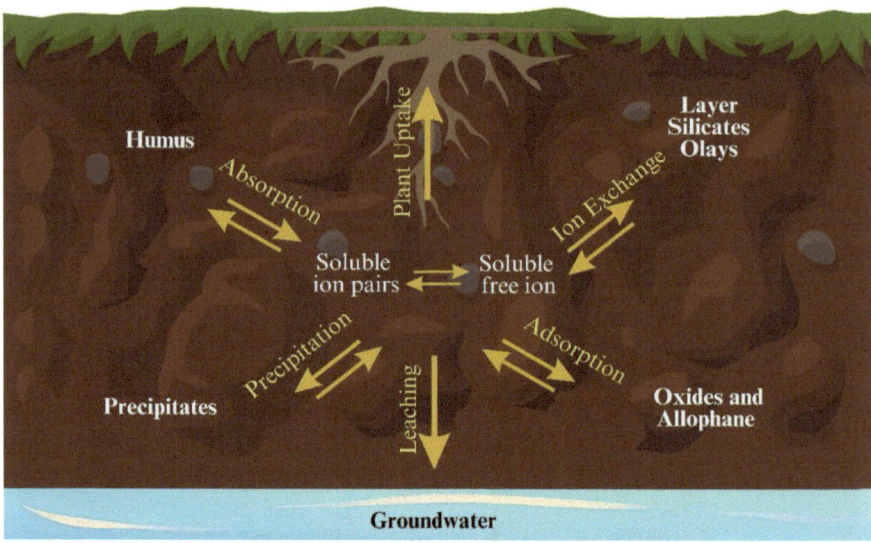

Fig. 1.5 Diagram illustrating dynamic interactive processes governing solubility, availability, and mobility of elements (Randive 2013)

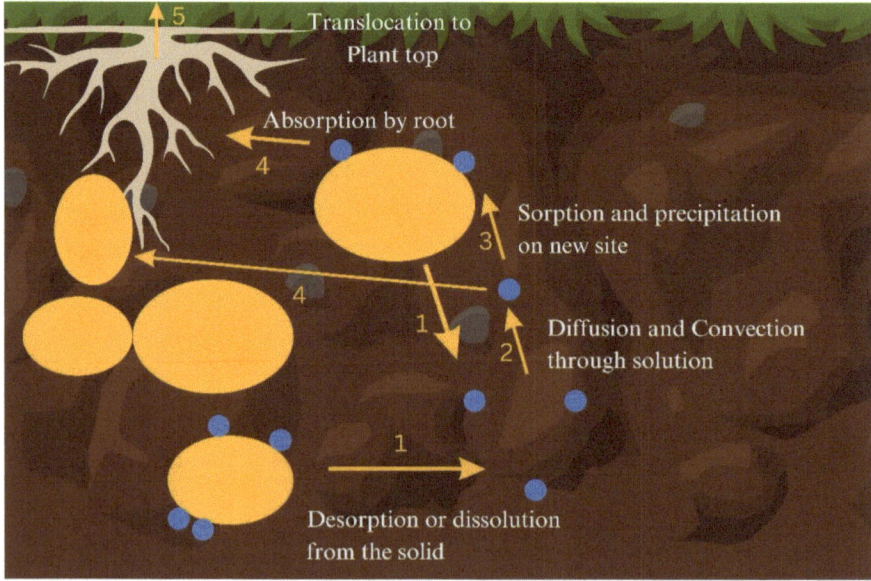

Fig. 1.6 Diagram illustrating stepwise intake of micronutrients by plants from soils (Randive 2013)

as limestone, sulfates, and phosphates, along with metals. For example, traditional Indian remedies include Swarnabhasma (gold powder), Rajatbhasma (silver powder) and Sheelajit (black bitumen). However, much of this ancient knowledge faded away with the rise of modern medicine. Chinese texts also describe use of earth materials for treatment of various ailments. Today, the pharmaceutical industry continues to use many geological materials from rocks and minerals for making medicines. Geopharmacy, which encompasses study of geological materials for their use in medicine, is an integral component of medical geology.

1.8.1 Internal Applications

Geological materials are important for treating internal health problems. For example, calcium from fossil shales of molluscs is used in medicines that help strengthen bones, which is especially beneficial for children (Mailafiya et al. 2019). Fluorite powder is an ingredient in toothpaste, helping to prevent dental issues (Vranić et al. 2004). Sulfa-drugs are made from sulfurous salts found in evaporites. Similarly, multivitamin tablets often contain metals like gold, silver, zinc, copper, and cobalt (Randive 2013).

1.8.2 External Applications

Clay minerals are important for external therapies (Nomicisio et al. 2023). Mud therapy, which involves applying mud packs to the skin, is well-known for treating various skin issues and other ailments. Beauty salons frequently use Bentonite and other clays in facial treatments. Limestone is essential for making plaster of Paris, that is commonly used to set bone fractures. Furthermore, soaking in hot springs, geysers, and solfataras is a popular method for treating many health problems. Balneotherapy, or bathing in thermal mineral waters, has been practiced for centuries, especially for chronic skin and musculoskeletal conditions (Huang et al. 2018).

References

Ahmed et al (2017) A review on rock cycle. Int J Emerg Technol Innov Res 4(12):100–102 (www. jetir.org), ISSN:2349-5162. Available: http://www.jetir.org/papers/JETIR1712021.pdf

Asiminicesei D-M, Fertu DI, Gavrilescu M (2024) Impact of heavy metal pollution in the environment on the metabolic profile of medicinal plants and their therapeutic potential. Plants 13(6):913. https://doi.org/10.3390/plants13060913

Azarm-Karnagh S, Zahrani F, Negat N (2020) An overview of the water cycle, its pollutants and threats

Bashir I, Lone FA, Bhat RA, Mir SA, Dar ZA, Dar SA (2020) Concerns and threats of contamination on aquatic ecosystems. Bioremed Biotechnol 27:1–26. https://doi.org/10.1007/978-3-030-356 91-0_1. PMCID: PMC7121614

Byrne JP, Barbara S (2012) CA; ABC-CLIO, LLCL: Encyclopedia of the black death, 429 pp

Chojnacka K, Mikulewicz M (2024) Bioaccumulation. In: Wexler P (ed) Encyclopedia of toxicology, 4th edn. Academic Press, pp 77–84, ISBN 9780323854344. https://doi.org/10.1016/B978-0-12-824315-2.00351-1

Finkelman RB, Centeno JA, Selinus O (2005) The emerging medical and geological association. Trans Am Clin Climatol Assoc 116:155–165; discussion 165. PMID: 16555612; PMCID: PMC1473139

Gitano (2024) The book of healing | PDF. Scribd. https://www.scribd.com/document/617910795/The-Book-of-Healing

Godbole P et al (2024) Innovative approach to transform mining waste into value added products. In: Randive K, Nandi AK, Jain PK, Jawadand S (eds) Current trends in mineral-based products and utilization of wastes: recent studies from India. MBD 2022. In: Springer proceedings in earth and environmental sciences. Springer, Cham. https://doi.org/10.1007/978-3-031-50262-0_18

Goldhaber M, Banwart S (2015) Soil formation. https://doi.org/10.1079/9781780645322.0082

Hasan SE, Finkelman RB, Skinner HCW (2013) In: The impact of the geological sciences on society. In: Bickford ME (ed) Geology and health: a brief history from the Pleistocene to today. Geological Society of America; Boulder, CO pp 155–164. Special Paper 501

Hasan SE (2019) Emerging trends in global health security and medical geology, short course in medical geology. In: MEDGEO 2019, 8th international meeting of the international medical geology association, Guiyang, China

Hasan SE (2021) Medical geology. Encycl Geol 684–702. https://doi.org/10.1016/B978-0-12-409 548-9.12523-0. Epub 2020 Dec 2. PMCID: PMC7241403

Hou D, O'Connor D (2020) Chapter 1—green and sustainable remediation: concepts, principles, and pertaining research. In: Hou D (ed) Sustainable remediation of contaminated soil and groundwater, Butterworth-Heinemann, pp 1–17, ISBN 9780128179826. https://doi.org/10.1016/B978-0-12-817982-6.00001-X

Huang A, Seité S, Adar T (2018) The use of balneotherapy in dermatology. Clin Dermatol 36(3):363–368. https://doi.org/10.1016/j.clindermatol.2018.03.010. Epub 2018 Mar 13. PMID: 29908578

Huang T, Hu Q, Shen Y, Anglés A, Fernández-Remolar DC (2024) Biogeochemical cycles. In: Scheiner SM (ed) Encyclopedia of biodiversity, 3rd edn. Academic Press, pp 393–407, ISBN 9780323984348. https://doi.org/10.1016/B978-0-12-822562-2.00347-9

Klaes B, Wörner G, Thiele-Bruhn S, Arz HW, Struck J, Dellwig O, Groschopf N, Lorenz M, Wagner J-F, Urrea OB, Lamy F, Kilian R (2022) Element mobility related to rock weathering and soil formation at the westward side of the southernmost Patagonian Andes. Sci Total Environ. https://doi.org/10.1016/j.scitotenv.2022.152977

Kumar S, Saha N, Mohana AA et al (2024) Atmospheric particulate matter and associated trace elements pollution in Bangladesh: a comparative study with global megacities. Water Air Soil Pollut 235:222. https://doi.org/10.1007/s11270-024-07021-8

Lag J (1990) Geomedicine. CRC Press, Boca Raton, FL (English translation) 288 pp

Muhammad Mailafiya M, Abubakar K, Danmaigoro A, Chiroma SM, Rahim EBA, Moklas MAM, Zakaria ZAB (2019) Cockle shell-derived calcium carbonate (aragonite) nanoparticles: a dynamite to nanomedicine. Appl Sci 9(14): 2897. https://doi.org/10.3390/app9142897

Medical geology (2024). Geological survey Ireland. https://www.gsi.ie/en-ie/geoscience-topics/env ironmental-health/Pages/Medical-geology.aspx

National Research Council (US) (2000a) Committee on copper in drinking water. copper in drinking water. National Academies Press (US), Washington (DC). Health effects of excess copper. Available from: https://www.ncbi.nlm.nih.gov/books/NBK225400/

National Research Council (US) (2000b) Commission on engineering and technical systems; National Research Council (US) Commission on life sciences. In: McKone TE, Huey BM,

Downing E et al (eds) Strategies to protect the health of deployed U.S. forces: detecting, characterizing, and documenting exposures. National Academies Press (US), Washington (DC). Environmental and exposure pathways. Available from: https://www.ncbi.nlm.nih.gov/books/NBK225345/

Nomicisio C, Ruggeri M, Bianchi E, Vigani B, Valentino C, Aguzzi C, Viseras C, Rossi S, Sandri G (2023) Natural and synthetic clay minerals in the pharmaceutical and biomedical fields. Pharmaceutics 15(5):1368. https://doi.org/10.3390/pharmaceutics15051368.PMID:37242610; PMCID:PMC10220772

Nzengung V, Gugolz S (2022) 14—Biochar-based constructed wetland for contaminants removal from manure wastewater. Mohan D, Pittman CU, Mlsna TE (eds) Sustainable biochar for water and wastewater treatment. Elsevier, pp 487–525, ISBN 9780128222256. https://doi.org/10.1016/B978-0-12-822225-6.00004-X

Randive K, Jawadand S, Sheikh A, Dora ML, Satyanarayanan M, Subramanyam KSV (2022) We did not notice this demon in our backyard!—introducing a new source of geogenic health hazard. In: Randive KR, Pingle S, Agnihotri A (eds) Medical geology in mining-health hazards due to metal toxicity. Springer Nature, pp 319–348

Randive KR, Godbole P, Jawadand S, Chopra V, Dora ML, Dhoble SJ (2023) 12-Radioactive waste management in India: present status and future perspectives. In: Raut NA, Kokare DM, Bhanvase BA, Randive KR, Dhoble SJ (eds) 360-degree waste management, vol 2. Elsevier, pp 273–298, ISBN 9780323909099. https://doi.org/10.1016/B978-0-323-90909-9.00005-8

Randive K (2013) Elements of geochemistry, geochemical exploration and medical geology

Ríos-Reyes CA (2021) The importance of minerals in medical geology: impacts of the environment on health. redalyc.org. https://www.redalyc.org/journal/2738/273865670016/html/

Sadhra AS (2013) (Amar) says medieval surgery: abulcasis. From the hands of quacks. https://fromthehandsofquacks.com/2013/02/19/medieval-surgery-abulcasis/

Selinus O, Finkelman RB, Centeno JA (2019) Principles of medical geology☆. Nriagu J (eds) Encyclopedia of environmental health, 2nd edn. Elsevier, pp 364–371, ISBN 9780444639523. https://doi.org/10.1016/B978-0-12-409548-9.11715-4

Selinus OS (2007) Medical geology: an opportunity for the future ambio. 36:114–116. https://doi.org/10.1579/0044-7447(2007)36[114:MGAOFT]2.0.CO;2

Selinus O (2020) Department of biology and environmental science. Linneaus University, Kalmar. Personal communications dated September 21, 2019 and Jan. 27, 2020

Selinus OS, Finkelman R, Centeno J, Lax K (2005) Medical geology: a new future for geoscience. Eur Geol 10:27–30

Share (2024a) Environmental mutagens and gene expression I learn science at scitable. nature.com. http://www.nature.com/scitable/topicpage/environmental-mutagens-cell-signalling-and-dna-repair-1090

Share (2024b) Soil minerals and plant nutrition I learn science at scitable. nature.com. https://www.nature.com/scitable/knowledge/library/soil-minerals-and-plant-nutrition-127881474/?error=cookies_not_supported&code=50474fc4-5247-4391-9f89-eb247f6cae32

Sunitha V, Reddy M (2012) Medical geology: a globally emerging discipline. J Adv Chem Sci

Tumane R, Pingle S, Jawade A, Randive K (2022) Toxicity and occupational health hazards of coal fly ash. In: Randive K, Pingle S, Agnihotri A (eds) Medical geology in mining. Springer geology. Springer, Cham. https://doi.org/10.1007/978-3-030-99495-2

Vranić E, Lacević A, Mehmedagić A, Uzunović A (2004) Formulation ingredients for toothpastes and mouthwashes. Bosn J Basic Med Sci 4(4):51–58. https://doi.org/10.17305/bjbms.2004.3362. PMID: 15628997; PMCID: PMC7245492

Wang F, Leilei X, Leung K, Elsner M, Zhang Y, Guo Y, Pan B, Sun H, An T, Ying G, Brooks B, Hou D, Helbling D, Sun J, Qiu H, Vogel T, Zhang W, Gao Y, Simpson M, Tiedje J (2024) Emerging contaminants: a one health perspective. The Innovation 100612. https://doi.org/10.1016/j.xinn.2024.100612

Chapter 2
Trace Elements and Human Health

Trace elements are crucial for maintaining human health, as they are involved in many important biological functions. This chapter highlights the role of trace elements, such as copper, zinc, selenium, and molybdenum, which are required in small amounts but are essential for enzyme activity, metabolism, and overall body functions. It explains how these elements enter the body, primarily through food and inhalation, and how they are absorbed and distributed. The chapter also categorizes and classifies these elements, discussing their importance and potential risks. Additionally, it covers recommended intake levels and how the body regulates these elements to maintain balance, emphasizing the need to understand these processes for better health management and to prevent deficiencies.

2.1 What Are Trace Elements?

Trace elements, also called trace metals, are known as minerals by medical professionals. They are present in very small amounts in living tissues. Some trace elements are essential for nutrition, while others are considered essential based on limited evidence. Most trace elements are non-essential, meaning the body does not need them (National Research Council 1989). These elements are classified as trace elements when they are found in the human body in quantities of milligrams per kilogram (mg/kg) of body weight or less. The term "ultra-trace element" refers to elements that are needed in even smaller amounts, usually less than 1 milligram per kilogram (mg/kg) and often below 50 micrograms per kilogram (μg/kg) for laboratory animals. For humans, this means an element needed in less than 1 milligram per day (mg/day), often measured in micrograms per day (μg/day) (Nielsen 2003). Trace elements are essential for many biological processes (the basic activities that keep the body working) (Spears and Engle 2016). They are important inorganic substances

(substances which do not contain carbon) needed in small amounts along with vitamins and other micronutrients. These elements are part of enzymes (proteins that speed up chemical reactions) and proteins that help cell function. If the body doesn't get enough of these trace elements, it can lead to illness, and in severe cases, it may cause death. However, grouping minerals into major, trace, or ultra-trace categories is not always clear-cut (Tsuji et al. 2016). It's important to get the right amount of trace elements, as too little or too much can cause problems. Even though they make up only 5% of a typical diet, minerals are vital for maintaining health. For essential elements, whether they are helpful or harmful depends on how much is consumed (Mehri 2020). A healthy body needs 60 elements, 15 vitamins, 12 essential amino acids (the building blocks of proteins), and 3 essential fatty acids (important fats that the body can't make on its own) every day. Dr. Linus Pauling, a Nobel Prize winner, linked many diseases to the lack of these essential nutrients. He believed that many illnesses could be prevented by getting the right nutrition, helping the body stay healthy and fight off sickness (Randive 2013).

2.2 Potential Pathways of Trace Elements to Enter Human Body

Trace elements enter the human body through multiple pathways, often originating from the Earth's crust and influenced by both natural processes and human activities (Li and Wu 2022). One of the main ways we are exposed to trace elements is through our diet. Plants absorb these elements from the soil via their roots (Andresen et al. 2018). As plants grow, they store these elements in their tissues, which are then consumed directly by humans or indirectly through animal products (Chojnacka and Mikulewicz 2014). Another significant route of exposure is through inhalation (Witkowska et al. 2021). Tiny particles in the air, carrying trace elements, can be inhaled into the lungs. These particles can originate from natural events like volcanic eruptions or dust storms, or from human activities such as industrial pollution, vehicle emissions, and farming (Kumar et al. 2024; Briffa et al. 2020). Once inside the body, trace elements follow a series of processes collectively referred to as pharmacokinetics (i.e., absorption, distribution, metabolism, and excretion (ADME) process). This involves their absorption into the bloodstream, distribution to various parts of the body, metabolism (where they are used or broken down), and excretion (how they are removed from the body) (Li et al. 2019). Elements like iron, calcium, and zinc are essential for many bodily functions, including supporting enzymes and facilitating biochemical reactions which are crucial for health (Jomova et al. 2022). However, when these elements are consumed in excess or when non-essential elements accumulate, they can become toxic and cause health problems. This makes both, understanding how these elements enter the body and maintaining their balance vital for health (Mehri 2020). To effectively manage both the benefits and potential risks of

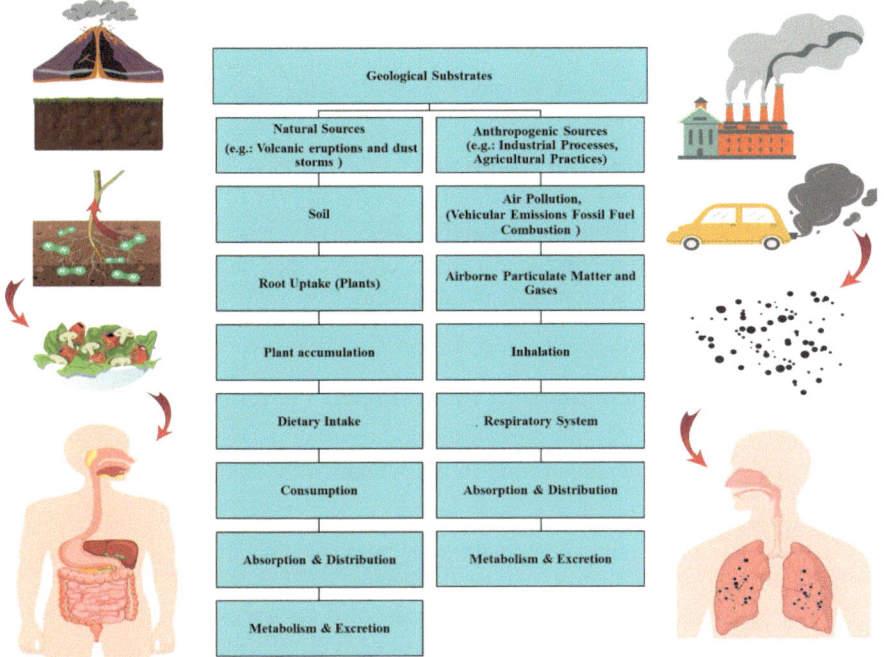

Fig. 2.1 Pathways for trace element transport in human body (modified from National Research Council 1989; Andresen et al. 2018; Chojnacka and Mikulewicz 2014; Witkowska et al. 2021; Kumar et al. 2024; Briffa et al. 2020; Li et al. 2019; Mehri 2020)

trace elements, it is important to categorize them based on whether they are essential or toxic, ensuring a clearer understanding of their role in human health (Fig. 2.1).

2.3 Classification of Trace Elements

Trace elements are essential in small amounts for our bodies to function properly. Classifying these elements helps us understand their specific roles in health and nutrition. Generally, trace elements are divided into essential and non-essential categories. Essential trace elements are necessary for our well-being, while non-essential ones can be harmful when present in excess. By grouping trace elements this way, we can better understand how important they are for our diets and the potential health problems that can arise from having too much or too little of them. Some of the fundamental classifications are discussed below.

2.3.1 WHO Classification, 1973 (WHO 1973)

According to this classification, nineteen trace elements are categorized into three groups:

1. **Essential elements**: zinc (Zn), copper (Cu), selenium (Se), chromium (Cr), cobalt (Co), iodine (I), manganese (Mn), and molybdenum (Mo).
2. **Probably essential elements**: nickel (Ni), silicon (Si), vanadium (V) and tin (Sn)
3. **Potentially toxic elements**: lead (Pb), mercury (Hg), cadmium (Cd), arsenic (As), thallium (Tl), barium (Ba), and antimony (Sb).

2.3.2 Frieden's Classification of Elements (Frieden 1974)

Twenty-nine types of elements found in the human body are categorized into five major groups as follows:

1. **Group I**: Basic components of macromolecules such as carbohydrates, proteins, and lipids. Examples include carbon, hydrogen, oxygen, and nitrogen.
2. **Group II**: Nutritionally important minerals, also known as principal or macro-elements. These macro-elements are required daily in amounts exceeding 100 mg/day for adults. Examples include sodium, potassium, calcium, phosphorus, magnesium, and sulphur.
3. **Group III**: Essential trace elements, also known as minor elements. These elements are considered trace elements when their daily requirement is below 100 mg. While deficiencies are rare, they can be fatal. Examples include copper, iron, zinc, chromium, cobalt, iodine, molybdenum, and selenium.
4. **Group IV**: Additional trace elements, whose roles are still unclear but may be essential. Examples include cadmium, nickel, silica, tin, vanadium, and aluminium. This group may correspond to the 'probably essential' trace elements in the WHO classification.
5. **Group V**: Non-essential metals with unknown functions that can be toxic in excess amounts. Examples include arsenic, cadmium, mercury, and lead. This group aligns with the potentially toxic elements defined in the WHO classification.

2.3.3 Frieden's Classification of Elements (Frieden 1985)

In 1981, Frieden introduced a biological classification of trace elements based on their concentration in tissues

1. **Essential trace elements**: boron, cobalt, copper, iodine, iron, manganese, molybdenum, and zinc.

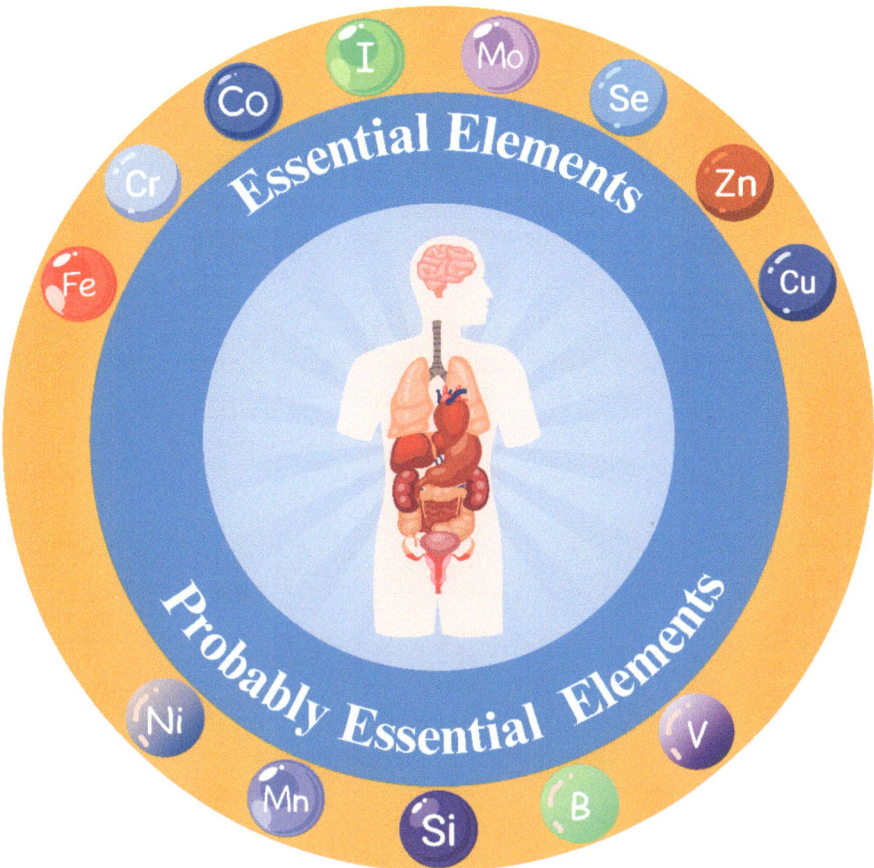

Fig. 2.2 Showing various types of essential and probably essential elements

2. **Probably essential trace elements**: chromium, fluorine, nickel, selenium, and vanadium.
3. **Physically promotive trace elements**: bromine, lithium, silicon, tin, and titanium (Fig. 2.2).

2.4 Recommended Dietary Allowances for Trace Element

The Recommended Dietary Allowances (RDA) set guidelines for how much of each essential trace element we should consume for good health. However, many people do not get even 75% of the recommended amounts for various trace elements. Recent studies have also raised questions about whether boron is truly essential for human health. Trace elements can affect each other, and having an imbalance can lead to health problems. Several factors, such as diet, how well our bodies absorb these

Table 2.1 Table displaying the estimated average requirement, recommended dietary allowances, and tolerable intake levels of common trace elements (source: Intakes for Vitamin A, Vitamin K, Arsenic, Boron, Chromium, Copper, Iodine, Iron, Manganese, Molybdenum, Nickel, Silicon, Vanadium, and Zinc) (Source: Randive 2013)

Nutrient	Estimated average requirements	Recommended dietary allowances	Tolerance upper intake levels	Unit
Boron	NE	–	20	mg
Calcium	NE	1000	2500	mg
Chloride	NE	2300	3600	mg
Chromium	NE	35	ND	ug
Copper	700	900	10000	ug
Fluoride	NE	4	10	mg
Iodine	95	150	1100	ug
Iron	6	8	45	mg
Magnesium	330	420	350	mg
Manganese	NE	7.1	11	mg
Molybdenum	34	45	2000	pg
Phosphorus	580	700	4000	mg
Potassium	NE	4700	ND	mg
Sodium	NE	1500	2300	mg
Sulfate	NE	–	ND	–
Zinc	9.4	11	40	mg

elements, toxic effects, and interactions with medications, can impact this balance. Because of these complexities, it's advised not to add arsenic, silicon, and vanadium to foods or supplements (Randive 2013) (Table 2.1).

2.5 Essential Trace Elements

2.5.1 Copper

Copper (Cu) is a trace element found primarily in minerals such as chalcopyrite, bornite, and malachite in the Earth's crust. It is extracted from these ores through mining activities. When copper is mined and processed, it can enter the environment and eventually reach our bodies. Copper enters the human body mainly through oral intake. Copper is absorbed in the digestive system and then utilized for various physiological functions. Copper is essential for the function of various enzyme and metabolism. The body requires about 2–5 mg of copper daily. It plays a critical role in the production of haemoglobin, which carries oxygen in the blood, and supports the body's energy metabolism. Copper deficiency can occur due to genetic or environmental factors. Deficiency of copper can lead to anaemia (a condition where the blood has fewer red blood cells) and neurological problems. On the flip side, excessive copper intake can cause nausea, vomiting, and liver dysfunction. Disorders like Wilson's disease lead to copper accumulation in the body, while Menkes syndrome results in insufficient copper. Copper levels can also be affected by conditions such as heart attacks and infections, which may weaken the immune system and lead to bone issues. Elevated copper levels are also associated with oral health problems, including cancer (Chitturi et al. 2015; Mehri, 2020; Al-Fartusie and Mohssan 2017).

2.5.2 Zinc

Zinc (Zn) is a vital trace element that plays numerous roles in the body. It is primarily found in minerals like sphalerite (zinc sulphide) and zincite (zinc oxide), which are extracted through mining activities. These minerals can release zinc into the environment, and humans absorb zinc mainly from dietary sources. Zinc is absorbed in the digestive system and utilized for various essential functions. Zinc serves many critical roles in the body, including acting as a catalyst to speed up chemical reactions. It also provides structural support to proteins, which are essential molecules that perform a wide range of functions in the body. Additionally, zinc regulates several biological processes, including maintaining homeostasis (the balance of bodily functions) and supporting immune responses (the body's defense against infections). It also helps manage oxidative stress (damage caused by free radicals), controls apoptosis (programmed cell death), and influences gene expression (how genes are turned on or off). Proteins that bind zinc, known as metallothioneins (MTs), help protect the body from stress, toxic metals, infections, and low zinc levels, supporting cellular balance. Adequate zinc intake improves immune function, reduces the risk of infections, and can help combat the effects of aging. However, excessive zinc intake can be harmful, leading to reduced absorption of other nutrients, increased zinc excretion (the removal of excess zinc through urine), and potential poisoning. Zinc deficiency, which is a lack of sufficient zinc in the body, is a global concern and can result from poor dietary habits, medical conditions, genetic factors, or treatments. Severe deficiency can cause symptoms such as skin rashes (dermatitis), hair loss (alopecia), weight loss, weakened immunity, low testosterone levels (hypogonadism), and slow wound healing. While zinc supplements are useful for addressing deficiencies, taking high doses over long periods can lead to anaemia (a condition where the body does not have enough red blood cells), emphasizing the need for proper usage (Mehri 2020; Stefanidou et al. 2006; Mohammad and Vadstrup 2014).

2.5.3   Selenium

Selenium (Se) is a trace element with a unique history, as it was once considered toxic but is now recognized as essential for our health. It is mainly found in minerals like selenite and selenate, which are often obtained from mining. People get selenium primarily from their diet, especially from foods such as Brazil nuts, seafood, grains, and meat. Selenium plays a crucial role in our bodies as a component of certain proteins and acts as a powerful antioxidant, protecting cells from damage by harmful molecules known as free radicals. It is involved in several important processes, including cancer development, supporting the immune system, male reproductive health, and cardiovascular health (the health of the heart and blood vessels). However, scientists are still studying how selenium specifically affects these processes. Recent research advances, such as bioinformatics (using computer tools to analyse biological data), cDNA microarrays (a technique for studying gene expression), and transgenesis (introducing new genes into organisms), have helped us understand how selenium influences various biological functions. Selenium is linked to Keshan disease (a heart disease caused due to selenium deficiency), and may help protect against viral infections like SARS-CoV-2, highlighting its importance for health. However, selenium has a narrow therapeutic window, meaning both too little and too much selenium can cause health problems. This makes it important to carefully manage selenium intake through diet and supplements.

2.5.4   Molybdenum

Molybdenum (Mo) is an essential trace element that plays a key role in various bodily functions. It can be found in various foods, including legumes, dairy products, and meats. The geological sources of molybdenum include minerals like molybdenite, which is the primary ore from which molybdenum is extracted. These minerals are often found in granitic rocks and certain sedimentary deposits, and can be mined for their molybdenum content. Molybdenum is a crucial part of the molybdenum cofactor complex, which helps certain enzymes work. These enzymes, such as xanthine oxidase (XO), aldehyde oxidase (AO), and sulfite oxidase (SO), are important for processes such as breaking down purines (these are chemical compounds that form uric acid when metabolized), converting aldehydes (a type of organic compound), and processing amino acids. While molybdenum deficiency is uncommon in typical diets, it can occur in people receiving long-term total parenteral nutrition (a method of feeding that bypasses the usual digestive process). Symptoms of deficiency can include a fast heart rate (tachycardia) and mental disturbances. Biochemical signs of molybdenum deficiency include reduced levels of uric acid and inorganic sulphate, which indicate that XO and SO are not working properly. When there are isolated deficiencies in these enzymes or in the molybdenum cofactor, serious issues can arise, such as intellectual disabilities and dislocation of the eye lens, caused by sulfite buildup and not enough sulphate for brain health. Moreover, molybdenum deficiency may be linked to higher rates of oesophageal cancer in areas where the soil lacks this element.

2.5.5 Iodine

Iodine (I) is a trace element that primarily occurs in seaweed, certain marine organisms, and some mineral deposits. Coastal areas and regions near oceans typically have higher iodine levels in the soil and water, making seafood an important dietary source of this essential element. Iodine is crucial for producing thyroid hormones, specifically thyroxine (T4, a hormone that regulates metabolism) and triiodothyronine (T3, a hormone that also plays a role in metabolism and growth). These hormones regulate growth, development, and many metabolic processes in the body. Iodine from our diet is absorbed as iodide (the ionic form of iodine), mainly by the thyroid gland, with any excess being removed from the body through urine. Thyroid hormones are especially important for the development of the brain and central nervous system of the infant during pregnancy and early childhood (up to three years old). They also assist in controlling how body uses carbohydrates, fats, proteins, vitamins, and minerals. A deficiency in iodine can lead to several health issues, including hypothyroidism (a condition where the thyroid does not produce enough hormones), which can cause fatigue, weight gain, and sensitivity to cold; goitre (an enlarged thyroid gland that can lead to swelling in the neck); congenital abnormalities (birth defects due to maternal iodine deficiency during pregnancy); increased risk of miscarriage and premature birth; cognitive impairments, including intellectual disabilities and developmental delays in children; neonatal death (the death of a baby within the first 28 days of life); and decreased blood sodium levels. In some regions where iodine deficiency is common, iodizing table salt has been implemented as a public health measure to prevent these health issues. The recommended daily intake of iodine for adults is 150 micrograms (μg). On the other hand, too much iodine can also cause health problems. It can lead to hyperthyroidism, where the thyroid makes too much hormone. This can cause weight loss, a fast heartbeat, and anxiety. It may also cause goitre, like when there is not enough iodine. Other issues include tachycardia (a very fast heart rate) and a higher risk of thyroid cancer. It is important to note that selenium plays a supportive role in using iodine effectively because it is needed for converting T4 to T3. Therefore, both iodine deficiency and excess can be detrimental, so balanced iodine intake is necessary (Paz et al. 2018; Mehri 2020).

2.5.6 Cobalt

Cobalt (Co) is geologically sourced from minerals such as cobaltite, erythrite, and skutterudite, and it is often obtained as a byproduct of nickel and copper mining.

Once released into the environment through mining and industrial processes, cobalt can enter the food chain through soil and water. Plants absorb cobalt from the soil, which is then ingested by humans through food consumption, especially animal products. In the human body, cobalt primarily functions as part of vitamin B12 (cyanocobalamin), which is vital for red blood cell formation and neurological health. In industrial settings, exposure to cobalt occurs through inhalation of dust or direct skin contact, particularly in sectors like hard metal manufacturing, cobalt refining, and alloy production. This exposure can lead to respiratory issues such as asthma and allergic dermatitis (skin inflammation). Prolonged ingestion of cobalt, often through contaminated food or water, can also result in cardiomyopathy (a disease of the heart muscle). To monitor cobalt exposure, urinary cobalt concentrations are often measured, especially in occupational settings, since cobalt has a short half-life in the body (Lauwerys 1994; MHRA 2017).

2.5.7 Chromium

Chromium (Cr) is a mineral found in various geological sources, particularly in chromite ore. It enters the human body mainly through the diet. The most common food sources of chromium are meat, grains, fruits, and vegetables. Traditionally, chromium has been considered an essential nutrient because it helps with glucose metabolism. This is the process by which human body processes sugar by enhancing the effects of insulin (a hormone that helps regulate blood sugar levels). However, recent studies have questioned the efficacy of chromium supplements in improving insulin sensitivity and lowering the risk of diabetes. Insulin sensitivity refers to how well the body responds to insulin; higher sensitivity means the body can better manage blood sugar levels. The exact process of how chromium is absorbed in the intestine is not fully understood. Its absorption can be influenced by other substances in the diet, such as oxalates (found in some plants), iron, and high sugar intake. Although chromium deficiency is rare, it has been noted in hospitalized patients who have higher metabolic needs. Interestingly, studies have shown that very low-chromium diets do not significantly affect how well the body tolerates glucose. The amount of chromium the body absorbs is minimal, and this absorption can vary based on how food is processed. While trivalent chromium (Cr III), which is a common form of chromium found in supplements and foods, is generally considered safer than the hexavalent chromium (Cr VI), which is highly toxic and linked to serious health issues such as lung cancer, particularly among workers exposed to it (Yoshida 2012; Mehri 2020).

2.5.8 Iron

Iron (Fe) is the most abundant essential trace element in the human body, totalling about 3–5 grams, primarily found in the blood, liver, bone marrow, and muscles, mainly in a form known as heme. Heme is a crucial component of haemoglobin (the protein in red blood cells that carries oxygen) and myoglobin (a protein that stores oxygen in muscles). Iron is commonly found in minerals like hematite and magnetite, which are mined from the Earth's crust and used to produce iron and steel. Iron can also be sourced from iron-rich foods such as red meat, poultry, fish, lentils, beans, and fortified cereals. When the body needs more iron, it absorbs it from food. The absorbed iron is then transported in the bloodstream as ferritin (a protein that stores iron). Hemosiderin is a byproduct formed when ferritin is broken down. The body uniquely regulates how much iron it absorbs to keep iron levels in the blood steady rather than excreting it. Iron exists in two forms: ferrous (Fe^{2+}) and ferric (Fe^{3+}). Both forms play significant roles in various enzymes involved in metabolic processes and energy production, including cytochrome a–c and cytochrome P450, which are important for many chemical reactions in the body. Iron deficiency can lead to several health issues, particularly related to oral health. For instance, iron deficiency anaemia occurs when there is not enough iron to produce haemoglobin, leading to symptoms like fatigue, a smooth and painful tongue (atrophic glossitis), and inflammation in the mouth (stomatitis). Another condition is Plummer-Vinson syndrome, a disorder marked by anaemia due to iron deficiency and the formation of web-like membranes in the throat, making swallowing difficult. This condition can increase the risk of cancer due to the formation of abnormal growths (webs) in the oesophagus. Oral conditions like oral submucous fibrosis (OSMF), which are linked to low serum iron levels, may also occur due to decreased collagen production. Increased consumption of areca nut, a common chewable substance in some cultures, can worsen this condition by affecting how iron is utilized in the body. In patients with head and neck cancers, elevated levels of ferritin (indicating iron stores in the body) and low serum iron levels can be observed, suggesting that monitoring iron levels may help manage these patients effectively. In summary, iron plays a vital role in many physiological processes, impacting both overall health and the progression of various diseases, especially those related to oral health (Vasudevan and Shreekumari 2007; Frieden 1972; Satyanarayana 2008; Bhattacharya et al. 2016).

2.6 Probably Essential Trace Elements

2.6.1 Manganese

Manganese (Mn) is an essential trace element that plays a vital role in human health. It is found in minerals such as pyrolusite and rhodochrosite, which can be sourced from mining activities and soil. Manganese enters the human body mainly through food, including nuts, whole grains, leafy vegetables, and teas. Manganese is crucial for the production of certain enzymes and helps regulate how the body uses glucose (a type of sugar) and lipids (fats). It is a key part of manganese superoxide dismutase (MnSOD), an enzyme that helps neutralize harmful molecules called reactive oxygen species (ROS), reducing oxidative stress in mitochondria (the energy-producing parts of cells). Both manganese deficiency and excessive exposure can lead to health problems, including metabolic disorders. Conditions like type 2 diabetes, obesity, insulin resistance, atherosclerosis (hardening of the arteries), hyperlipidemia (high levels of fats in the blood), non-alcoholic fatty liver disease (NAFLD), and hepatic steatosis (fat accumulation in the liver) have been linked to increased ROS generation and inflammation. A lack of manganese or too much exposure can worsen these issues by raising ROS levels and thereby oxidative stress. Manganese is stored in bones and is also found in significant amounts in the liver, pancreas, kidneys, and brain. However, measuring manganese levels accurately in the body can be difficult due to limited clinical methods. The Dietary Reference Intakes (DRIs) offer guidelines for manganese intake, which vary by age and gender. The Adequate Intake (AI) levels range from 600 micrograms per day for infants to 2.6 milligrams per day for lactating women. These AI levels help ensure adequate manganese intake for optimal health and metabolism (Li and Yang 2018; Mehri 2020).

2.6.2 Silicon

Silicon (Si) has become recognized as an essential trace element important for the normal metabolism of higher animals. It is mainly found in minerals like quartz, and enters the human body through dietary sources, such as whole grains, fruits, vegetables, and beverages like beer. Silicon plays a key role in connective tissue, particularly in bones and cartilage. It helps with the formation of collagen (a protein that provides structure to tissues) and glycosaminoglycans (molecules that support tissue health). A deficiency in silicon can cause problems in these tissues, mainly affecting the creation of the organic matrix (the part of the tissue that gives it structure) more than the mineralization (the addition of minerals). In the body, silicon is abundant in osteogenic cells (bone-forming cells), particularly when they are actively functioning. It is also found in high concentrations within their mitochondria, which are the energy centres of the cell. Silicon contributes to the functions of subcellular structures that contain enzymes (proteins that help speed up chemical reactions) and interacts significantly with other elements. Beyond its metabolic roles, silicon is thought to have a structural function in connective tissue. Its connection to aging may be linked to changes in glycosaminoglycans, which are important for maintaining healthy connective tissues (Carlisle 1986; Mehri 2020).

2.6.3 Nickel

Nickel (Ni) is an essential trace element found in various geological sources, including minerals such as pentlandite, garnierite, and nickel laterites. These minerals are commonly mined from the Earth's crust. Nickel can enter the human body primarily through dietary sources like nuts, seeds, whole grains, and certain seafood. It is also present in small amounts in drinking water and air, especially near industrial areas. In humans and animals, about 1–10% of dietary nickel is absorbed, distributing across various organs such as the kidneys, bones, and lungs, where it can accumulate in significant concentrations. In the bloodstream, nickel primarily binds to a protein called albumin, along with smaller molecules like amino acids and peptides. The body eliminates nickel mainly through urine, but it can also be excreted through hair, skin, milk, and sweat. Although researchers are still trying to understand nickel's specific biological functions in humans, it is known to be important in various enzymatic processes, especially in microorganisms where nickel-containing enzymes play vital roles. Interestingly, no cases of nickel deficiency have been reported in humans, suggesting that taking nickel supplements may not be necessary. On the other hand, too much exposure to nickel—especially in workplaces or from wearing nickel-plated items like jewellery—can lead to health problems. These issues may include skin allergies, lung disease, kidney damage, and cardiovascular problems. While nickel is essential in animals like sheep, goats, pigs, and rats—where its deficiency can lead to growth issues and reproductive problems—its exact role in human health is still being researched (Kosák and Ubreza 2012; Datt et al. 2023; Mehri 2020).

2.6.4 Boron

Boron (B) is a trace element found in various geological sources, primarily in the form of borate minerals like borax and kernite. It can enter the human body through dietary sources, as it is naturally present in fruits, vegetables, nuts, and legumes. Once consumed, boron is absorbed in the gastrointestinal tract and excreted through urine. Boron plays a role in several important bodily functions, including steroid hormone metabolism, bone development, and the maintenance of cell membranes. It is primarily stored in bones, nails, and hair. While the World Health Organization (WHO) states that the essentiality of boron for humans is still uncertain, its presence in various biological processes suggests it may have important health benefits. Although boron is generally safe at normal dietary levels, excessive intake can lead to developmental issues. Occasional exposure to boron typically does not result in toxicity, but long-term exposure may cause neurological problems and kidney damage. A deficiency in boron can negatively affect growth, bone development, and cognitive functions and has been linked to Kashin–Beck disease, a bone disorder found in certain regions of China. Boron also influences metabolic processes, including how the body uses macro-minerals and energy sources. The body regulates boron levels, indicating a desirable daily intake of over 1 mg but not exceeding 13 mg. Diets lacking in fruits, vegetables, legumes, and nuts may not provide enough boron, highlighting its potential nutritional importance that deserves further recognition (Pizzorno 2015; Nielsen 1997; Mehri 2020).

2.6.5 Vanadium

Vanadium (V) is a trace element found in the environment, especially in minerals like vanadinite and carnotite. It is present in soil, water, and many foods. Humans absorb vanadium mainly through (breathing) and mouth (eating). The typical daily intake ranges from 32.6 to 135 micrograms, but most of this is excreted in feces, which means the body usually doesn't accumulate it. While vanadium is generally safe at these levels, exposure to high amounts, especially from industrial sources, can be harmful. Cases of vanadium deficiency in humans are rare, but some studies suggest that low intake may be linked to heart diseases. Research has shown that vanadium compounds might help in treating conditions like type 2 diabetes, cancer, and bacterial infections, but they are not yet approved for medical use. Vanadium likely helps to regulate processes that depend on phosphate in the body. In the bloodstream, vanadium spreads to various tissues, with bones acting as a storage site. In water, vanadate is common at lower concentrations, while forms like decavanadate are more stable at higher pH levels. Although vanadium is widely studied, it has not been confirmed as essential for humans, but it is recognized as important for some animals, like goats. More research is needed to understand its role in human health (Rehder 2013; Harland and Williams 1994) (Table 2.2).

2.7 Absorption of Trace Elements in Body

The absorption of trace elements in the human body is a crucial process that ensures the effective uptake of these essential nutrients from the diet for various physiological functions. This absorption primarily occurs in the gastrointestinal tract, particularly in the duodenum and jejunum (the distal two parts of the small intestine), where optimal pH facilitates uptake (National Research Council 1989). Specialized carrier systems exist for specific trace elements, such as iron and zinc, aiding their transport across the intestinal wall (mucosa) (Kiela and Ghishan 2016). For example, iron is absorbed via a carrier-mediated process involving divalent metal transporter 1 (DMT1) and ferroportin (a transmembrane protein that transports iron from the inside of a cell to the outside of the cell). The chemical form of trace elements also influences absorption, with organic forms or chelated compounds generally being more efficiently absorbed than inorganic forms. Factors like dietary composition, the presence of other nutrients, and physiological status can impact trace element

Table 2.2 Overview of essential and potentially essential trace elements: health benefits, toxicity, natural and geological sources

Element	Classification	Health benefits	Excess toxicity	Natural source	Geological source	References
Copper (Cu)	Essential	Supports iron metabolism, immune function, and collagen formation	Can cause liver damage and gastrointestinal issues	Shellfish, nuts, whole grain	Chalcopyrite, malachite	Mehri (2020)
Zinc (Zn)	Essential	Essential for immune response, wound healing, and DNA synthesis	Can cause nausea and disrupt copper absorption	Meat, nuts, beans	Sphalerite	Mehri (2020), Stefanidou et al. (2006)
Selenium (Se)	Essential	Acts as an antioxidant, supports thyroid function, and boosts immunity	Can lead to hair loss and gastrointestinal distress	Brazil nuts, seafood	Selenite	Mehri (2020), Saito (2022)
Molybdenum (Mo)	Essential	Important for enzyme function and metabolism of sulphur-containing amino acids	May cause joint pain	Legumes, grains	Molybdenite	Mehri (2020)
Iodine (I)	Essential	Crucial to produce thyroid hormones, regulating metabolism	Excess can lead to thyroid dysfunction	Seafood, iodized salt	Iodine-rich minerals	Mehri (2020)
Cobalt (Co)	Essential	Vital for red blood cell formation and maintaining nerve health	High levels can affect heart health	Meat, dairy	Cobaltite	Mehri (2020)

(continued)

Table 2.2 (continued)

Element	Classification	Health benefits	Excess toxicity	Natural source	Geological source	References
Chromium (Cr)	Essential	Helps regulate blood sugar levels and supports insulin function	Excess may harm kidneys	Meat, whole grains	Chromite	Mehri (2020)
Iron (Fe)	Essential	Essential for oxygen transport in blood and energy metabolism	Excess can cause organ damage	Red meat, beans	Hematite	Mehri (2020)
Manganese (Mn)	Potentially essential	Supports bone health, metabolism, and antioxidant defence	High levels may lead to neurological issues	Nuts, whole grains	Pyrolusite	Mehri (2020), Li and Yang (2018)
Silicon (Si)	Potentially essential	Supports bone formation, skin health, and connective tissue	Rarely causes toxicity	Fruits, vegetables	Silicates	Carlisle (1986), Mehri (2020)
Nickel (Ni)	Potentially essential	May support enzyme function and cardiovascular health	Can cause skin allergies	Nuts, grains, chocolate	Nickel laterite	Mehri (2020)
Boron (B)	Potentially essential	Supports bone health, hormone balance, and cognitive function	Excess can lead to developmental issues	Fruits, nuts, vegetables	Borate minerals	Pizzorno (2015), Mehri (2020)
Vanadium (V)	Potentially essential	May help improve blood sugar control and support cholesterol metabolism	High levels can affect respiratory health	Seafood, mushrooms	Vanadinite	Rehder (2013), Mehri (2020)

absorption. For instance, vitamin C enhances iron absorption by reducing ferric iron to ferrous iron, whereas, certain dietary fibers or phytates can inhibit absorption by forming insoluble complexes with trace elements (Piskin et al. 2022a, 2022b). Furthermore, the body regulates trace element homeostasis through feedback mechanisms that adjust absorption rates based on needs and excretion processes that eliminate excess trace elements (Maares and Haase 2020). Hormonal regulation, such as the role of calcitriol (active vitamin D) in calcium absorption, also influences trace element uptake (Institute of Medicine 2011). In summary, the absorption of trace elements is a highly regulated process involving specialized mechanisms influenced by various factors, including chemical form, dietary composition, and physiological status. Understanding these processes is crucial for optimizing trace element intake and effectively addressing deficiencies.

References

Al-Fartusie F, Mohssan S (2017) Essential trace elements and their vital roles in human body. Ind J Adv Chem Sci 5:127–136. https://doi.org/10.22607/IJACS.2017.503003

Andresen E, Edgar P, Hendrik K (2018) Trace metal metabolism in plants. J Exp Bot 69(5):909–954. https://doi.org/10.1093/jxb/erx465

Bhattacharya PT, Misra SR, Hussain M (2016) Nutritional aspects of essential trace elements in oral health, and disease: an extensive review. Scientifica (Cairo) 2016:5464373. https://doi.org/10.1155/2016/5464373. Epub 2016 Jun 28. PMID: 27433374; PMCID: PMC4940574

Briffa J, Sinagra E, Blundell R (2020) Heavy metal pollution in the environment and their toxicological effects on humans. Heliyon 6(9):e04691, ISSN 2405-8440. https://doi.org/10.1016/j.heliyon.2020.e04691

Carlisle EM (1986) Silicon as an essential trace element in animal nutrition. Ciba Found Symp. 121:123–39. https://doi.org/10.1002/9780470513323.ch8. PMID: 3743227

Chojnacka K, Mikulewicz M (2014) Bioaccumulation. In: Wexler P (ed) Encyclopedia of toxicology, 3rd edn. Academic Press, pp 456–460, ISBN 9780123864550. https://doi.org/10.1016/B978-0-12-386454-3.01039-3

Chitturi R, Baddam V, Prasad LK, Lingamaneni P, Kattapagari K (2015) A review on role of essential trace elements in health and disease. J Dr. NTR Univ Health Sci 4. https://doi.org/10.4103/2277-8632.158577

Datt SC, Thamizhan P, Chauhan P, Dudi K, Mani V (2023) Effects of nickel supplementation on nutrient utilization, mineral balance, haematology and antioxidant status of crossbred dairy calves. J Trace Elem Med Biol 79:127250, ISSN 0946-672X. https://doi.org/10.1016/j.jtemb.2023.127250

Frieden E (1947) The evolution of metals as essential elements [with special reference to iron and copper]. In: Friedman M (ed) Protein-metal interactions, vol 48. Springer, New York, NY, USA, pp 1–31 (Advances in Experimental Medicine and Biology)

Frieden E (1972) The chemical elements of life. Sci Am 227(1):52–60

Frieden E (1985) New perspectives on the essential trace elements. J Chem Educ 62(11):915–923. https://doi.org/10.1021/ed062p915

Harland BF, Harden-Williams BA (1994) Is vanadium of human nutritional importance yet? J Am Diet Assoc. 94(8):891–894. https://doi.org/10.1016/0002-8223(94)92371-x. PMID: 8046184

Institute of Medicine (US) (2011) Committee to review dietary reference intakes for vitamin D and calcium; Ross AC, Taylor CL, Yaktine AL et al (eds) Dietary reference intakes for calcium and vitamin D. Overview of calcium, vol 2. National Academies Press (US), Washington (DC). Available from: https://www.ncbi.nlm.nih.gov/books/NBK56060/

Jomova K, Makova M, Alomar SY, Alwasel SH, Nepovimova E, Kuca K, Rhodes CJ, Valko M (2022) Essential metals in health and disease. Chem Biol Interact 367:110173, ISSN 0009-2797. https://doi.org/10.1016/j.cbi.2022.110173

Kumar S, Saha N, Mohana AA et al (2024) Atmospheric particulate matter and associated trace elements pollution in Bangladesh: a comparative study with global megacities. Water Air Soil Pollut 235:222. https://doi.org/10.1007/s11270-024-07021-8

Košǎk U, Obreza A (2012) Nickel as important trace element? Farmacevtski Vestnik. 63:297–304

Kiela PR, Ghishan FK (2016) Physiology of intestinal absorption and secretion. Best Pract Res Clin Gastroenterol 30(2):145–159. https://doi.org/10.1016/j.bpg.2016.02.007. Epub 2016 Feb 10. PMID: 27086882; PMCID: PMC4956471

Lauwerys R (1994) Biological monitoring of workers exposed to cobalt metal, salt, oxides and hard metal dust. Occup Environm Med 51:447–50

Li L, Yang X (2018) The essential element manganese, oxidative stress, and metabolic diseases: links and interactions. Oxid Med Cell Longev 2018:7580707. https://doi.org/10.1155/2018/758 0707. PMID: 29849912; PMCID: PMC5907490

Li P, Wu J (2022) Medical geology and medical geochemistry: an editorial introduction. Expo Health 14(2):217–218. https://doi.org/10.1007/s12403-022-00479-z. Epub 2022 Apr 22. PMID: 35474720; PMCID: PMC9025997

Li Y, Meng Q, Yang M, Liu D, Hou X, Tang L, Wang X, Lyu Y, Chen X, Liu K, Yu AM, Zuo Z, Bi H (2019) Current trends in drug metabolism and pharmacokinetics. Acta Pharm Sin B 9(6):1113–1144. https://doi.org/10.1016/j.apsb.2019.10.001. Epub 2019 Oct 18. PMID: 31867160; PMCID: PMC6900561

Mehri A (2020) Trace elements in human nutrition (II)—an update. Int J Prev Med 11(1):2. https://doi.org/10.4103/ijpvm.IJPVM_48_19

Muhamed PK, Vadstrup S (2014) [Zinc is the most important trace element]. Ugeskr Laeger 176(5):V11120654. Danish. PMID: 25096007

MDA/2017/018—all metal-on-metal (MoM) hip replacements: updated advice for follow-up of patients (Issued by MHRA 29 June 2017)

Maares M, Haase H (2020) A guide to human zinc absorption: general overview and recent advances of in vitro intestinal models. Nutrients 12(3):762. https://doi.org/10.3390/nu12030762. PMID: 32183116; PMCID: PMC7146416

National Research Council (US) (1989) Committee on diet and health. Diet and health: implications for reducing chronic disease risk. Trace elements, vol 14. National Academies Press (US), Washington (DC). Available from: https://www.ncbi.nlm.nih.gov/books/NBK218751/

Nielsen FH (1997) Boron in human and animal nutrition. Plant and Soil 193:199–208. https://doi.org/10.1023/A:1004276311956

Nielsen FH (2003) Trace elements. Caballero B (ed) Encyclopaedia of food sciences and nutrition, 2nd edn. Academic Press, pp 5820–5828, ISBN 9780122270550. https://doi.org/10.1016/B0-12-227055-X/01204-9

Paz S, Rubio C, Gutiérrez A, Revert C (2018) Iodine: an essential trace element

Pizzorno L (2015) Nothing boring about boron. Integr Med (Encinitas) 14(4):35–48. PMID: 26770156; PMCID: PMC4712861

Piskin E, Cianciosi D, Gulec S, Tomas M, Capanoglu E (2022a) ACS Omega 7(24):20441–20456. https://doi.org/10.1021/acsomega.2c01833

Piskin E, Cianciosi D, Gulec S, Tomas M, Capanoglu E (2022b) Iron absorption: factors, limitations, and improvement methods. ACS Omega 7(24):20441–20456. https://doi.org/10.1021/acsomega.2c01833. PMID: 35755397; PMCID: PMC9219084

Randive K (2013) Elements of geochemistry, geochemical exploration and medical geology

Rehder D (2013) Vanadium. Its role for humans. Met Ions Life Sci 13:139–69. https://doi.org/10.1007/978-94-007-7500-8_5. PMID: 24470091; PMCID: PMC7120733

Satyanarayana U, Chakrapani U (2008) Essentials of biochemistry, 2nd edn. Book and Allied, Kolkata, India

Saito Y (2022) Essential trace element selenium and redox regulation: its metabolism, physiological function, and related diseases. Redox Exp Med (1):R149–R158. https://doi.org/10.1530/REM-22-0010

Stefanidou M, Maravelias C, Dona A, Spiliopoulou C (2006) Zinc: a multipurpose trace element. Arch Toxicol 80(1):1–9. https://doi.org/10.1007/s00204-005-0009-5. Epub 2005 Sep 27. PMID: 16187101

Spears JW, Engle TE (2016) Feed ingredients: feed supplements: microminerals, reference module in food science. Elsevier, ISBN 9780081005965. https://doi.org/10.1016/B978-0-08-100596-5.00760-5

Tsuji PA, Canter JA, Rosso LE (2016) Trace minerals and trace elements. Caballero B, Finglas PM, Toldrá F (eds) Encyclopaedia of food and health. Academic Press, pp 331–338, ISBN 9780123849533. https://doi.org/10.1016/B978-0-12-384947-2.00699-1

Vasudevan DM, Sreekumari S (2007) Text book of biochemistry for medical students, 5th edn. Jaypee, New Delhi, India

WHO (1973) Trace-elements in human nutrition. Report of a WHO Expert Committee. World Health Organization, Geneva, Switzerland (WHO Technical Report Series, No. 532)

Witkowska D, Słowik J, Chilicka K (2021) Heavy metals and human health: possible exposure pathways and the competition for protein binding sites. Molecules 26(19):6060. https://doi.org/10.3390/molecules26196060. PMID: 34641604; PMCID: PMC8511997

Yoshida M (2012) [Is chromium an essential trace element in human nutrition?]. Nihon Eiseigaku Zasshi 67(4):485–491. Japanese. https://doi.org/10.1265/jjh.67.485. PMID: 23095360

Chapter 3
Toxic Traits of Trace Elements

This chapter explores the impact of trace elements on human health, emphasizing their sources, exposure pathways, and harmful effects. It begins by outlining both natural and human-made sources of trace elements and how they enter the environment. The focus is on non-essential trace elements, which are not required by the body and may cause problems. The chapter examines various ways in which humans can be exposed to these elements. Similarly, how they are absorbed, transported, stored, and eventually eliminated from the body. It also explains how these elements become toxic and the potential health risks they pose. Overall, this chapter provides an overview of the toxic effects of trace elements and highlights their significance for human health.

3.1 Toxicity of Trace Elements

The term 'toxicity' refers to how harmful a substance is to humans or animals, depending on the duration and amount of exposure. Acute toxicity occurs when damage happens after a short-term or single exposure, while sub-chronic toxicity results from continuous exposure over a period longer than a year but shorter than an organism's lifespan. Chronic toxicity, on the other hand, stems from long-term exposure, often throughout an individual's life (Stoppler 2024). Even essential trace elements, which are required for normal bodily functions, can become harmful if consumed in excessive amounts over time. The body's tolerance to these elements determines safe intake levels, and exceeding those limits can lead to adverse health effects. Both essential trace elements such as molybdenum (Mo), copper (Cu), zinc (Zn), nickel (Ni), manganese (Mn), cobalt (Co), selenium (Se), and chromium (Cr), as well as non-essential or harmful elements like zirconium (Zr), antimony (Sb), arsenic

K. Randive and P. Godbole, *Medical Geology for Beginners*,
SpringerBriefs in Medical Earth Sciences, https://doi.org/10.1007/978-3-031-82765-5_3

(As), lead (Pb), mercury (Hg), and cadmium (Cd), can severely disrupt the biochemical, structural, and functional processes in living organisms (National Research Council 1989; Antoniadis et al. 2019).

Some trace elements are classified as heavy metals due to their high density (greater than 5 g/cm^3) and metallic properties. These include cadmium (Cd), cobalt (Co), chromium (Cr), copper (Cu), iron (Fe), manganese (Mn), molybdenum (Mo), nickel (Ni), lead (Pb), tin (Sn), vanadium (V), and zinc (Zn) (Pourret and Bollinger 2018; Wu et al. 2018). Additionally, metalloids like arsenic (As) and antimony (Sb), as well as non-metals like selenium (Se), can be harmful, even though they are not classified as heavy metals. While heavy metals are naturally present in the Earth's crust, human activities such as mining, smelting, industrial production, and the use of metals in agriculture and households are the main causes of environmental contamination. These metals enter the environment through processes such as metal corrosion, atmospheric deposition, soil erosion, and leaching into soil and water. Contamination can also result from the resuspension of sediments in water and soil, and the evaporation of metals into the air.

Natural processes like weathering and volcanic eruptions contribute to trace elements' pollution, but human activities are the primary source. Industrial sources include metal processing, coal combustion in power plants, petroleum burning, nuclear power production, and industries such as plastics, textiles, microelectronics, wood preservation, and paper production (Tchounwou et al. 2012). Trace elements enter the environment through both natural and man-made activities (Liu et al. 2018; Rinklebe et al. 2019). Industrial practices, mining, agriculture, and waste disposal contribute to the accumulation of these elements in soil. When exposure levels exceed critical thresholds, whether through soil, air, water, or the food chain; these elements can negatively affect the biochemical, structural, and functional systems of microbes, plants, animals, and humans (Antoniadis et al. 2019).

3.2 Entry and Transport of Trace Elements

Local Toxicity

Toxic elements can impact the human body in two main ways: directly at the entry point, known as local toxicity, or through absorption into the body. Local toxicity refers to harmful effects that occur right where a toxic substance enters the body, such as the lungs through inhalation. This type of toxicity typically occurs before the substance is absorbed into the bloodstream. However, under particular circumstances, local effects are also shown after absorption, if they occur at the site of entry (for example, localized nerve damage or injury). Local toxicity occurs when a toxic agent is caustic or corrosive, causing burns or damage upon contact. It can also arise from irritating chemicals that lead to inflammation. If present in high concentrations at the entry site, the agent may produce immediate, acute effects. For instance, when chlorine gas is inhaled, it can directly damage lung tissue, causing swelling and

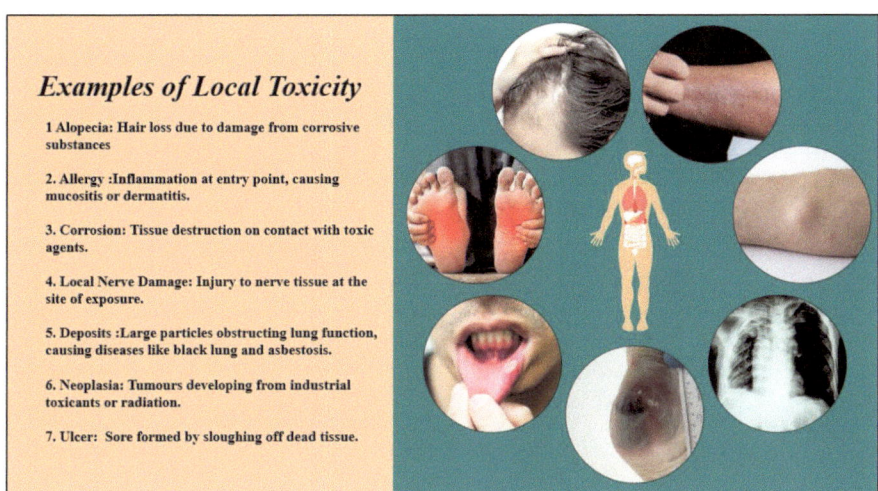

Fig. 3.1   An image illustrating examples of local toxicity (modified after Randive 2013)

harm. This illustrates local toxicity, as the harmful effects occur at the point of entry into the body. Some examples of local toxicity include skin burns from contact with caustic substances and respiratory distress from inhaling irritant gases (Randive 2013) (Fig. 3.1).

3.3   Mechanisms of Toxicity

The toxicity of trace metals depends on several factors, including how much of the metal is being ingested, how living organisms absorb it, and how long they are exposed to it. For a substance to be classified as a toxicant, it must meet certain criteria: it should interact with biological targets (the specific molecules it affects), disrupt their normal functions, reach effective amounts (sufficient to cause harm) at the target site, and lead to noticeable changes in the body. A metal becomes a toxicant when it interacts with important biological molecules such as the receptors (proteins on the surface of cells that help the cell respond to signals), the Enzymes (proteins that help speed up chemical reactions in the body), the DNA (The material that carries genetic information in living organisms), the proteins (large molecules that have many roles in the body, like building tissues and speeding up reactions) and the lipids (fats that are important for the structure and function of cells) (Mood et al. 2021; Alekseenko et al. 2024). When a toxicant interacts with these molecules, it disrupts their normal functions, leading to harmful effects on health. The main toxicant is usually the specific metal that the organisms encounter through the environment. The source of these metals are usually rocks, minerals or other anthropogenic substances that enter the food chain through water, air, soil, or plants. In rare cases, exposure to these toxic

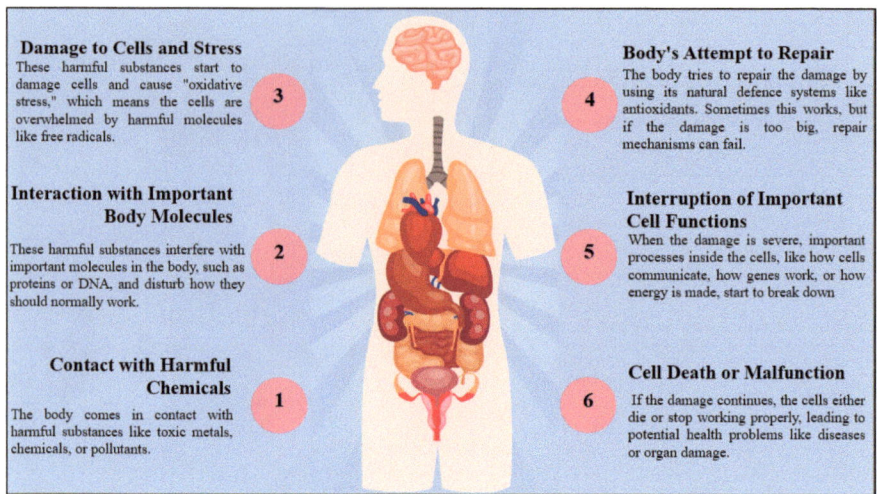

Fig. 3.2 Mechanism of toxicity: a step-by-step overview (modified after Shankar 2008)

metals forms the reactive oxygen species (ROS) or reactive nitrogen species (RNS) within the body, that are highly reactive and harms cells and tissues (Shanker 2008).

Trace metals can enter the body through various processes, accumulate in different areas, and cause harm. The absorption of these elements in the body depends on its geological source and pathway. As they enter the body, they are reabsorbed and activated, which increases their hazard potential. In simple words, the process of toxicity usually follows a sequence. It starts with metals being delivered from the environment to the body where they interact with the biological molecules. This interaction triggers chemical reactions which disrupts the normal body functions. The resulting damage is observed at different levels, from small molecules to entire systems (Shanker 2008; Langman and Kapur, 2006; Gregus and Klaassen 2001) (Fig. 3.2).

3.4 Absorption of Trace Element

When toxic agents enter the body, the process of crossing cell membranes at the entry point and entering the bloodstream is called absorption. After being absorbed, the toxic agent moves through the thin lining of the organ into the fluid between cells. Later, it spreads to target cells in the body, by the process called distribution. The way toxic substances move across cell membranes depends on their chemical structure. They can move through passive transport (without energy), active transport (using energy), or facilitated diffusion (using helpers like proteins). Different entry points can affect how well certain chemicals are absorbed (Lois and Mc-Keeman 2024). The main areas where absorption takes place include the following:

3.4.1 Gastro-Intestinal (GI) Tract

Toxic substances are not easily absorbed in the mouth and oesophagus because they pass through these areas quickly. The stomach, with its high acidity (pH 1–3), absorbs weak organic acids like benzoic acid, which are in a form that easily dissolves in fats (lipid-soluble). On the other hand, weak bases are mostly ionized and therefore poorly absorbed in the stomach. The small intestine is the main site for absorption due to its large surface area and near-neutral pH (5–8). Both weak acids and weak bases are absorbed easily here through passive diffusion. However, strong acids, strong bases, large molecules, and metals like lead and thallium are absorbed via active transport, which requires energy (Alsanosi 2014).

3.4.2 Lungs

In the lungs, most absorption takes place in the pulmonary region, which includes tiny airways called bronchioles and the alveolar sacs. These sacs have very thin membranes that separate inhaled air from the blood, allowing substances to pass through them easily. Efficacy of a toxic substance to absorption depends on its physical form—whether it is a gas, aerosol, or particulate matter. For gases, the absorption rate depends on their capacity to get dissolved in the blood. Highly soluble gases like chloroform enter the blood easily through passive diffusion. However, gases with low solubility, such as ethylene, quickly saturate the blood, stopping further absorption. For aerosols and particulate matter, the absorption depends on the size of the particles (Cavallazzi and Folz 2012).

3.4.3 Skin

In the skin, the epidermis is the main layer that controls the entry of toxic agents. The epidermis has seven layers, but the most important one is the outermost layer called the *stratum corneum*. This layer contains keratin, a tough protein that makes the cells more resistant to chemicals and harder to penetrate than the deeper layers. Toxic agents that get through the stratum corneum do so by passive diffusion. Water-soluble substances can pass through the outer surface of the hydrated keratin layer. Once the toxic substance passes through the stratum corneum, it can reach blood vessels located about 100 μm beneath the surface of the skin (Yu et al. 2021; Primak and Blumer 2006).

3.5   Distribution and Systematic Toxicity

Distribution refers to the movement of an absorbed toxic agent from the entry site to other parts of the body via blood or lymph. Blood transports chemicals faster than lymph, so most distribution occurs through the blood. All tissues and organs with a blood supply can be exposed to the absorbed chemical. For chemicals absorbed through the GI tract, the distribution follows a specific path. The toxic agent first flows into the portal system, where it is transported to the liver before passing through the heart and lungs. This allows the liver to immediately break down the toxic agent, and the lungs can eliminate it before it spreads further. The binding ability of a toxic agent to plasma proteins like albumin greatly affects its distribution. Only unbound chemicals can pass through capillary membranes, so the tighter the binding, the less it spreads in the body (Lois and Mc-Keeman 2024). Toxicology uses a measure called apparent volume of distribution (Vd), which is the dose of a toxin divided by its plasma concentration. A high Vd means more toxin is distributed to tissues, while a low Vd indicates less distribution (Davis et al. 2011).

$$\text{Apparent Volume of Distribution (Vd)} = \frac{\text{Dose of a Toxin (mg)}}{\text{Plasma Concentration of the Agent (mg/L)}}$$

The amount of a toxic agent that an organ or tissue receives depends on blood flow and the presence of barriers that slows down its entry. Organs with more blood flow, like the liver, kidneys, heart, and brain, are exposed to higher levels of toxins. Barriers such as the blood-brain barrier, placenta, and testes can slow down the absorption of toxic agents but do not completely block them. The distribution of a toxic agent also depends on how easily it can enter tissues (permeability) and its preference to certain tissues (affinity). For example, carbon monoxide binds to haemoglobin more strongly than oxygen, while fat-soluble toxins are attracted to fatty tissues (Slitt 2024).

Systemic Toxicity

Systemic toxicity occurs when a toxic substance spreads throughout the body after absorption, affecting various organs:

1. *Blood*: Blood is highly susceptible to poisoning. For example, carbon monoxide can bind to haemoglobin, preventing oxygen transport, and chloramphenicol can reduce white blood cell counts (ATSDR 2024a, b).
2. *Cardiovascular*: Toxins can damage the heart, potentially leading to alcoholic cardiomyopathy (Randive 2013).
3. *Dermal*: Internal distribution can lead to skin issues like phototoxicity from certain drugs and chloracne from chlorine-containing agents. Moreover, exposure to chlorine-containing agents has been linked to chloracne, a severe skin condition characterized by acne and skin inflammation (Sullivan et al. 2005).

4. *Endocrine*: Toxins can harm endocrine glands or interfere with hormone action. For example, aluminium can inhibit parathyroid hormone secretion (Sullivan et al. 2005).
5. *Hepato*: The liver is vulnerable to toxins due to its blood supply and role in detoxification. Effects include steatosis from carbon tetrachloride or ethanol and fatty liver from copper (Welcome to ToxTutor 2024).
6. *Immuno*: Immunotoxicity can occur without being localized, such as immunosuppression from cocaine and leukaemia from benzene (ATSDR 2024a, b).
7. *Nephro*: The kidneys are common toxicity sites because they filter blood. For example, lead can damage kidney cells (Welcome to ToxTutor 2024).
8. *Neuro*: Neurotoxicity affects the central and peripheral nervous systems. Heavy metals like mercury can cause demyelination (Welcome to ToxTutor 2024).
9. *Ocular*: Toxins in the bloodstream can reach the eyes, leading to damage like cataracts or optic nerve issues. For instance, methanol can harm the optic nerve (ATSDR 2024a, b).
10. *Reproductive (teratogenesis)*: Toxic agents can harm both male and female reproductive systems, affecting germ cells (sperm and ova) through DNA damage and other cellular injuries. For instance, lead accumulation in the testes can cause testicular degeneration and reduce sperm production (Randive 2013).

3.6 Storage and Excretion of Toxic Elements

Storage

The storage site in the human body is where substances are held in high concentrations for future use. Toxic agents can be stored in different areas based on their structure and how they bind. These agents may or may not cause harm at the storage site. They strive to maintain balance with the free form of the element in the blood. When a toxic substance is broken down or removed from the body, the concentration in the blood decreases, prompting more release from storage. Increased blood flow to a tissue can enhance the storage of a toxic agent, but if a specific toxic agent has a strong attraction to a tissue, it can accumulate there even with lower blood flow (Welcome to ToxTutor 2024). Major storage sites for toxic agents in the body are detailed below.

(a) *Plasma Protein*: Toxic agents can bind to plasma proteins during distribution in the blood, acting as storage sites based on the strength of the bond. These proteins can bind hydrophobic, hydrophilic, and neutral molecules. For instance, transferrin (main protein regulating Fe homeostasis) and ceruloplasmin (a circulating ferroxidase enzyme that is able to oxidize ferrous ions to less toxic ferric forms) bind iron and copper, while metallothionein (a protein involved in metal detoxification) binds cadmium in the kidneys and liver. The most abundant plasma protein, albumin, can bind various compounds, including metal ions

(Ca^{2+} and Zn^{2+}), acid dyes, barbiturates, and fatty acids due to its hydrophilic and hydrophobic properties (Plasma Protein Binding 2021).

(b) *Adipose Tissue*: Fat is primarily located in subcutaneous tissue (the deepest layer of skin) and serves as a storage site for lipophilic toxic agents that easily dissolve in body fat (Geyer et al. 1993). This storage reduces the concentration of toxic agents in target organs, meaning individuals with a higher body fat percentage may experience less intoxication. Adipose tissue continually exchanges lipids with the blood, allowing toxic agents to be released for distribution or redeposited in other fat cells. For example, during rapid fat mobilization, such as starvation, toxic agents like organochlorine insecticides can be released into the bloodstream and affect target organs (Jackson et al. 2017).

(c) Bone: Bone consists of proteins and the mineral hydroxyapatite. During bone formation, calcium and hydroxyl ions are incorporated into the hydroxyapatite matrix. Some elements can substitute for these ions, such as fluoride displacing hydroxyl ions and lead or strontium substituting for calcium. Chronic fluoride deposition can lead to skeletal fluorosis, while strontium may cause osteosarcoma. The deposition of toxic agents in bone is not irreversible. Through osteoclastic activity, bone tissue can be absorbed and toxic agents released into the bloodstream (Randive 2013).

(d) Liver and Kidney: These organs serve as important storage sites due to their high blood flow and the presence of binding proteins. For example, ligandin transports organic anions, while metallothionein binds cadmium and zinc in the liver and kidneys (Goering and Barber 2010).

Excretion

Eliminating toxic elements from the body is crucial for determining their potential toxicity. Rapid elimination generally indicates less capacity to cause damage. Transport to elimination sites typically occurs through the circulatory system and depends on the water solubility of the toxic agent. Water-soluble chemicals can easily exit the storage or damage site, dissolve in blood, and reach excretory organs. Major excretion routes include the kidneys, gastrointestinal (GI) tract, and respiratory systems, with less common routes such as cerebrospinal fluid (CSF), sweat, saliva, and milk (Schmid 2024).

(a) *Urine*: The kidneys primarily excrete toxic agents via urine. During filtration, dissolved toxic agents in blood diffuse into the nephron's glomerulus, with excretion depending on size and water solubility. Reabsorption, based on polarity, allows ionized substances to remain in urine while lipid-soluble agents can be reabsorbed into circulation, extending their half-life and potential toxicity (Blake and Rosenblum 2014).

(b) *Faeces*: Toxicants are eliminated in faeces through two processes: excretion in bile entering the intestine and direct excretion from the GI tract. The biliary route is significant for faecal excretion of certain heavy metals (e.g., arsenic, lead, mercury). Some substances may be reabsorbed and returned to the liver, but most bile-excreted substances are water-soluble and unlikely to be reabsorbed. However, intestinal flora can bio-transform these substances, making them lipophilic and prolonging their presence in the body through enterohepatic circulation i.e. the movement of bile acid molecules from the liver to the small intestine and back to the liver (Gregus 1986).

(c) *Exhalation*: The lungs serve as a route for excreting gaseous toxic agents. Low solubility gases (e.g., ethylene) are eliminated more rapidly than high solubility gases (e.g., chloroform). Volatile liquids like diethyl ether are also excreted by exhalation. This route is efficient for lipid-soluble substances in gas or volatile liquid form (Randive 2013).

(d) *Others*: Toxic agents can also be eliminated through various routes. Most effect among these is vomiting, which can expel unabsorbed toxic agents, often stimulated by emetics., Others include milk, which may contain toxic elements resembling calcium (e.g., lead), the cerebrospinal fluid can remove agents from the central nervous system. Similarly, heavy sweating can lead to the elimination of certain metals (e.g., cadmium, copper). Expulsion through hair and nails is another pathway for heavy metal removal, while excretion in tears, skin, and other minor routes is generally less significant (Schmid 2024).

3.7 Effects of Toxic Elements on the Human Body

Mechanisms of Injury

Toxic agents can harm the body through various mechanisms, causing local and systemic effects at the molecular level. They typically disrupt normal biochemical or cellular processes. The importance of the disrupted function helps determine how harmful the toxic agent is.

3.7.1 Disruption of Essential Components: This Can Occur Through Several Actions

Replacement of Substances: Toxic agents can replace vital substances in the body with harmful substitutes. For example, in carbon monoxide (CO) poisoning, CO binds to hemoglobin 210 times more effectively than oxygen, preventing oxygen transport to cells for respiration. In bones, fluoride can replace hydroxyl ions, and lead or strontium can substitute for calcium, making bones less flexible (An Introduction to Toxic Substances 2024).

Receptor Agonist/Antagonist: Some toxic agents can bind to specific cell receptors, causing effects like natural hormones (agonism) or blocking these effects (antagonism). For example, certain organochlorines can bind to estrogen receptors, mimicking estrogen's effects and increasing breast cancer risk (Endocrine Disruptors 2024).

Inhibition of Enzymes and Proteins: Toxic agents can bind to and block the activity of important enzymes and proteins. For instance, DDT can inhibit Na^+K^+ ATPase, disrupting nerve function. Cadmium and mercury can interfere with vitamin D activation, affecting calcium and magnesium metabolism (Randive 2013).

Damage to Structural Integrity: Some toxic agents, like ethanol, can embed themselves in cell membranes, changing their fluidity and disrupting protein function. For example, substances like cigarette smoke and alcohol can lead to an overproduction of reactive oxygen species (ROS), damaging cells and tissues (introtoxsubstances 2017).

3.7.2 Alteration or Incorrect Expression of Genetic Material

Toxic agents can damage DNA through mechanisms like strand breakage, oxidation, and mutations. They can also interfere with the processes of gene transcription and translation by binding to crucial elements such as transcription factors (Barnes et al. 2018; Jia et al. 2021). This can lead to several harmful outcomes:

Inability to Produce Functional Proteins: When toxic agents bind to DNA or cause mutations, it can result in incorrect nucleotide pairing during replication. If not repaired, these errors can lead to faulty protein production. For instance, mercury can bind to DNA, resulting in dysfunctional proteins in the brain and kidneys (Share 2024a, b).

Carcinogenesis: Carcinogenic agents can potentially harm anybody site, including the GI tract, skin, lungs, and organs involved in metabolism and elimination. Mutagenic agents, such as ionizing radiation and heavy metals (e.g., arsenic, chromium), can directly mutate DNA, increasing cancer risk (Radfard et al. 2023).

Mutagenesis in Germ Cells: Mutations in germ cells can lead to teratogenic effects. For example, acrylamide in processed foods can reduce male fertility (Pourentezari 2024).

3.7.3 Direct Cell Death Without Affecting Chemical Processes

Certain agents can cause cell death without altering biochemical processes, often through corrosive effects, radiation, or heat. Common types include:

Allergy: A hypersensitive immune reaction triggered by specific substances, resulting in inflammation. For example, contact dermatitis can occur from allergens like chromium or mercury (Randive 2013).

Necrosis: This involves the failure of essential cell components, causing cell swelling and death. Necrosis typically affects groups of adjacent cells or tissue (Randive 2013).

Apoptosis: A programmed cell death where the cell dismantles itself into fragments. Most toxicologists don't classify allergic reactions as toxic effects since they result from an immune response rather than direct toxicity (Randive 2013).

Metaplasia: This is the conversion of one mature cell type to another in response to chronic irritation, aiming to create a more resilient tissue. However, this often leads to a loss of the original cell's function (Randive 2013).

Fibrosis: After necrosis, if tissue regeneration is insufficient, it may be replaced by connective tissue, resulting in fibrosis. This reduces organ function and can lead to complete dysfunction (Randive 2013).

Neoplasia: This refers to the abnormal growth of tissue, commonly known as a tumour Neoplasia can be benign or malignant, with malignant tumours (cancer) depleting resources from surrounding cells and impairing their function (Randive 2013).

3.8   Dose Wise Response to Toxic Elements

It is often said that "all substances are poisons; only the right dose makes a remedy or a poison." This highlights the importance of dose in understanding toxic effects. Even the most dangerous chemicals may have no harmful effect, or even positive effects, at very low doses. On the other hand, even essential elements like oxygen and water can be harmful or deadly in very high amounts. The type of effect that occurs depends on the dose of the toxic substance, how often and how long is the exposure, and the way the toxic agent acts at a molecular level (Gooch 2023). These effects can be:

(a) **Acute**: Acute effects occur immediately after a single, high-dose exposure to a toxic agent. This can involve a highly potent substance or one that is taken in large amounts. When the toxic substance enters the body, it can cause damage either at the site of exposure (like the skin, lungs, or digestive tract) or in areas where it is absorbed (like the blood or brain) (Division of Research Safety 2024). The damage happens because the toxic substance disrupts cell function or causes cell death in the tissues it first contacts. Acute skin exposure to toxicants such as mercury can lead to contact dermatitis (Toxicity 2024).

(b) **Subchronic**: Subchronic effects occur from repeated exposure to smaller doses over several weeks. In this case, the toxic substance is absorbed, distributed to an organ, and usually either processed (bio-transformed) or excreted. The small dose may cause minor or undetectable damage initially. However, with repeated exposure, the damage becomes cumulative, leading to more serious problems in the affected organ over time. Long-term, low-level workplace exposure to lead can cause anaemia after several weeks (Deng et al. 2022).

(c) **Chronic**: Chronic effects are the result of repeated, long-term exposure to small doses of a toxic substance over months or years (longer than subchronic). Chronic exposure can lead to progressive damage to organs responsible for processing and eliminating the toxic agent, or it can cause cancer to develop. Both local effects (at the point of contact) and systemic effects (throughout the body) can occur. Workers exposed to lead for years can develop chronic kidney disease, and coal miners can suffer from pulmonary fibrosis (lung damage) Division of Research Safety 2024).

3.9   Trace Elements and Associated Health concerns

1. **Arsenic (As)**

Inorganic arsenic, a toxic substance, is commonly found in groundwater due to natural geological sources such as arsenopyrite and other arsenic-bearing minerals in rocks and soils. These minerals release arsenic into water, particularly in areas with mining activities or geothermal features like hot springs. Regions such as Bangladesh, India, and parts of the United States are known for significant arsenic contamination in drinking water. Inorganic arsenic is a known carcinogen and the predominant chemical contaminant in drinking water worldwide. Acute arsenic poisoning manifests with symptoms like vomiting, abdominal pain, and diarrhoea, leading to numbness, muscle cramps, and potentially death. Long-term exposure, primarily through drinking water and food, leads to skin changes, lesions, and hyperkeratosis, potentially evolving into skin cancer in a span of about five years. Moreover, arsenic exposure is linked to bladder and lung cancers, and its carcinogenicity is confirmed by the International Agency for Research on Cancer (IARC) (Farzan and Karagas 2013; Quansah et al. 2015). Chronic ingestion of arsenic may also result in developmental issues, diabetes, pulmonary and cardiovascular diseases, and increased mortality due to myocardial infarction. Additionally, adverse pregnancy outcomes and infant mortality are associated with arsenic exposure, with long-term impacts on cognitive development, intelligence, and memory, potentially leading to various health complications in adulthood (Tolins et al. 2014). Arsenic can directly impact the reproductive system by inducing oxidative stress, which in turn disrupts its normal function (Zargari et al. 2022).

2. **Cadmium (Cd)**

Cadmium (Cd) is primarily sourced from the weathering of base metal mineral deposits, especially those rich in zinc, lead, and copper. Geological processes can release cadmium into the environment, contaminating soil and water. Human activities, including mining, industrial processes, and the burning of fossil fuels, also contribute to cadmium exposure. Additionally, it can accumulate in food crops, particularly in rice and leafy vegetables, posing significant health risks when consumed. Cadmium is a toxic by-product of zinc production, leading to its accumulation in the human body. It primarily targets kidneys, causing renal dysfunction, and can result in bone demineralization either directly or through kidney impairment. Exposure to cadmium is linked to lung diseases like emphysema, asthma, and bronchitis, along with high blood pressure and an increased risk of various cancers, including those of the lung, breast, prostate, nasopharynx, pancreas, and kidney. The toxic effects on the liver and kidneys results into the binding of cadmium ions which induces oxidative stress, resulting in mitochondrial damage and increased production of reactive oxygen species (ROS) (Genchi et al. 2020a, b). Exposure to cadmium predominantly occurs through contaminated food and water, inhalation, and cigarette smoking. The FAO/WHO recommends a tolerable cadmium intake of 0.4–0.5 mg/week for adults. Non-smokers and individuals with non-occupational exposure primarily encounter cadmium through their diet, with absorption occurring via the respiratory (13–19%) and digestive systems (10–44%) (Charkiewicz et al. 2023). Chronic low-level cadmium exposure in industrialized countries has been associated with adverse kidney and bone health effects, as evidenced by links between renal and bone biomarkers and urinary cadmium excretion. Efforts to remediate cadmium pollution include phytoremediation using plants like sunflower (Helianthus annuus L.), Indian mustard (Brassica juncea), and river red gum (Eucalyptus camaldulensis), nanoparticle-based remediation with TiO_2 and Al_2O_3 nanoparticles, and microbial fermentation for removing cadmium from food (Genchi et al. 2020a, b). Continuous monitoring of individuals exposed to heavy metals is essential for maintaining public health and implementing effective preventive measures (Charkiewicz et al. 2023).

3. **Lead (Pb)**

Lead (Pb) is primarily sourced from sedimentary rocks like black shales and limestones, which can accumulate lead-bearing sediments. The main ore, galena (PbS), is a mineral composed of lead and sulfur, often found in hydrothermal veins and sedimentary deposits alongside other lead minerals like cerussite ($PbCO_3$) and anglesite ($PbSO_4$). Major mining areas include China, Australia, the United States, and Canada, with notable deposits in the Mississippi Valley and the Appalachian region. Increasing human-made contaminants have heavily polluted the atmosphere, posing significant global environmental risks. Lead, a highly toxic metal, it presents substantial health hazards when it settles on the earth's surface. Lead-induced toxicity, caused by oxidative stress. It can damage cell membranes, DNA, and disrupt antioxidant defence systems of the body. Lead exposure can affect various organs, including the lungs, blood vessels, brain, testes, and liver. Both acute high-dose and chronic low-dose lead exposure can produce neurotoxic effects, with severe cases leading to lead encephalopathy, a condition that causes symptoms like irritability, headaches, memory loss, and hallucinations (Mandal et al. 2023). Lead exposure can impair neuropsychological functioning throughout life and negatively impact reproductive health in both sexes, leading to decreased libido, chromosomal damage, and impaired spermatogenesis (the production of sperm). Lead can cross the placenta, affecting foetal development and increasing the risk of complications such as spontaneous abortion (the loss of pregnancy) (Mandal et al. 2023). Symptoms of lead poisoning can vary widely among individuals, with some showing no symptoms despite elevated lead levels in their blood (Kosnett 2005; Mycyk et al. 2005). The variability in lead toxicity requires further research (Dart et al. 2004). Organic lead, which is a form of lead that contains carbon, is more toxic than inorganic lead due to its ability to dissolve in fats (lipid solubility) (Timbrell 2008). Blood Pb levels between 25–60 µg/dL can cause neuropsychiatric effects, while levels above 50 µg/dL may lead to anaemia (a deficiency of red blood cells), and levels over 80 µg/dL can cause abdominal pain in adults (Kosnett 2005; Merill et al. 2007). Levels exceeding 100 µg/dL may lead to severe encephalopathy, with children experiencing serious symptoms at lower levels (~70 µg/dL) (Henritig 2006; Brunton et al. 2007). Chronic lead exposure can result in memory loss, depression, nausea, abdominal pain, and numbness, with additional symptoms like fatigue, sleep problems, headaches, and anaemia (Patrick 2006; Pearce 2007). Pregnant women with elevated blood lead levels risk premature birth and low birth weight babies, with the foetus affected at concentrations below 25 µg/dL (Bellinger et al. 1987). Neonates (newborns) often have higher blood lead levels than their mothers (Shannon 2003; Bellinger 2005). Children are particularly vulnerable to lead poisoning due to rapid growth, higher absorption rates of lead, and behaviours that increase exposure risk (Landrigan et al. 2002; Chisolm and Harrison 1956), leading to aggressive behaviour and developmental delays.

4. **Mercury (Hg)**

Mercury (Hg) is a toxic metal prevalent in the environment, primarily sourced from volcanic rocks, sedimentary deposits, and sulphide minerals. Natural processes such as weathering and volcanic eruptions release mercury, while human activities like mining and industrial emissions further contribute to its pollution. The main ore mineral for mercury is cinnabar (HgS), commonly found in regions like Spain and California. Mercury exists in three forms: elemental, inorganic, and organic. Inorganic mercury compounds are water-soluble and have a bioavailability of 7–15% after ingestion, leading primarily to kidney damage and gastrointestinal irritation, causing symptoms like abdominal pain and diarrhoea. Elemental mercury is mainly absorbed through inhalation, quickly reaching vital organs, especially the brain and kidneys, and is lipid-soluble, allowing it to cross the blood-brain barrier, resulting in neurological symptoms such as tremors, insomnia, memory loss, and cognitive dysfunction. In pregnant women, mercury can cross the placenta, posing significant risks to foetal neurodevelopment, leading to cognitive issues and developmental delays in newborns. Chronic exposure is linked to cardiovascular issues, including carotid atherosclerosis and myocardial infarction. Mercury also disrupts the endocrine and immune systems, affecting hormonal balance and immune responses. In men, exposure can cause decreased libido, chromosomal damage, and impaired spermatogenesis, while women face increased risks of pregnancy complications, such as spontaneous abortion. High consumption of fish, especially shark and swordfish, exacerbates these health risks (Zulaikhah 2020; Mahaffey 2005).

5. **Chromium (Cr) (particularly hexavalent chromium, Cr (VI))**

Chromium (VI) pollution is a significant global environmental issue primarily sourced from chromite ores found in ultramafic rocks. Commonly used in various industries, chromium (VI) is non-biodegradable, contaminating soil, water and plants, and posing serious health risks to humans and wildlife. It is known to be carcinogenic (cancer-causing), genotoxic (damaging DNA), and mutagenic (causing mutations). Current methods for removing chromium (VI) from the environment are often not eco-friendly and require multiple chemicals. However, some microbes can naturally reduce chromium toxicity through processes like biosorption, reduction, efflux, and bioaccumulation (Sharma et al. 2022). Chromium (VI) is recognized as a respiratory carcinogen, primarily affecting industrial workers through inhalation, and contributing to lung cancer. Although limited evidence exists regarding toxicity from ingestion, studies have shown increased cancer risks among populations in China exposed to chromium (VI) through drinking water. For the general population, food consumption is the primary exposure route. The California Environmental Protection Agency (CalEPA) has set a public health goal for total chromium in drinking water due to observed stomach tumours in mice. Workers in chromate production industries have a higher incidence of lung cancer and nasal irritation, further supporting the link between inhaled chromium (VI) and lung cancer (Behrens et al. 2023). Dermal exposure occurs through contact with consumer products containing chromium and contaminated soils, particularly affecting construction workers dealing with cement. Chromium can penetrate damaged skin, bind to immune cells, and lead to dermatitis and skin ulcers (Hagvall et al. 2021).

6. **Nickel (Ni)**

Nickel exposure can lead to various health issues, including allergies, cardiovascular and kidney diseases, lung fibrosis, and cancers of the lung and nasal passages. The primary geological sources of nickel include nickel-rich ores such as pentlandite and garnierite, typically found in ultramafic rocks and lateritic soils. Nickel can enter the environment through mining activities, industrial processes, and the combustion of fossil fuels, contaminating soil and water. While the exact molecular mechanisms of nickel toxicity are not fully understood, mitochondrial dysfunction and oxidative stress are believed to be key contributors (Genchi et al. 2020a, b). Mitochondrial dysfunction refers to the impairment of mitochondria, which are the cell's energy-producing structures. When they do not function properly, they can produce less energy and generate harmful by-products. Oxidative stress occurs when there is an imbalance between free radicals (unstable molecules that can damage cells) and antioxidants (substances that neutralize free radicals) in the body. Nickel is recognized as haematotoxic (affecting blood), immunotoxic (affecting the immune system), neurotoxic (affecting the nervous system), genotoxic (causing DNA damage), reproductive toxic (affecting reproduction), pulmonary toxic (affecting the lungs), nephrotoxic (affecting the kidneys), hepatotoxic (affecting the liver), and carcinogenic (cancer-causing). Upon exposure, nickel generates free radicals in various tissues, leading to DNA damage, lipid peroxidation (damage to cell membranes caused by free radicals), and disturbances in calcium and sulfhydryl balance. The primary mechanism of nickel toxicity involves the depletion of glutathione, which is an important antioxidant that protects cells from damage, and binding to sulfhydryl groups in proteins, which can disrupt their normal function. Maintaining proper nickel levels in the body, activating protective signalling pathways, and utilizing both enzymatic (produced by the body) and non-enzymatic (obtained from diet) antioxidants are crucial for countering nickel toxicity and its potential to cause cancer (Das et al. 2008).

7. **Thallium (Tl)**

Thallium, a toxic metal found in some sulphide ores, can enter the environment through mining activities, industrial processes, and the burning of fossil fuels. Geological sources include thallium-rich minerals like crookesite and tlalocite, typically associated with lead and zinc deposits. Acute exposure to thallium can cause various symptoms affecting the gastrointestinal, neurological, and skin systems. Initially, individuals may experience abdominal pain, nausea, vomiting, and diarrhoea. This can be followed by neurological symptoms such as peripheral neuropathy (nerve damage) and tremors. Ocular issues like double vision (diplopia) and drooping eyelids (ptosis) may also occur, along with skin-related symptoms ranging from general irritation to hair loss (alopecia) and Mees lines (which are white lines that appear on nails). Other potential symptoms include decreased sweating (hypohidrosis), inflammation of the tongue (glossitis), and cardiovascular issues like rapid heart rate (tachycardia). It's essential to conduct thorough physical examinations focusing on the abdomen, nervous system, and eyes (Kemnic and Coleman 2023). Chronic exposure can lead to lasting neurological deficits, persistent hair loss, and cardiovascular problems. Thallium disrupts enzyme function and energy production within cells, causing further cellular damage over time. Timely detection through urine tests and strict environmental regulations are vital to prevent accidental poisonings (Genchi et al. 2021).

8. **Aluminium (Al)**

Aluminium (Al^{3+}) is a metal commonly found in the earth's crust that can enter the environment through mining, industrial processes, and the use of aluminium products. Geological sources include bauxite ore, from which aluminium is extracted. Aluminium poses health concerns due to its ability to bind with proteins and its unknown physiological role in the human body. Neurotoxic effects are especially noted in dialysis patients who were given aluminium salts as phosphate binders in dialysate. High levels of aluminium in blood and brain tissue have been linked to symptoms such as confusion, memory problems, and dementia. The brain's slow removal of aluminium, combined with its effects on various biological processes, contributes to these health issues. Specifically, aluminium can cause oxidative stress, disrupt calcium signalling in the hippocampus (an area important for memory), and affect cholinergic neurons, which are crucial for the production of acetylcholine, a neurotransmitter involved in memory and learning. This has raised concerns about a potential connection between aluminium exposure and Alzheimer's disease. Although neurological changes have been observed in people with high aluminium exposure, severe conditions like encephalopathy (brain dysfunction) are rare. There is, however, one reported case of rapidly progressive encephalopathy possibly linked to aluminium-induced lung fibrosis (Klotz et al. 2017).

9. **Antimony (Sb)**

Antimony exposure presents a number of health risks, encompassing various diseases and health complications. Inhalation, ingestion, or skin contact with antimony can precipitate chronic health issues, including irritation of the eyes, skin, and respiratory system. Prolonged exposure may exacerbate gastrointestinal (related to the stomach and intestines) disorders, alter electrocardiograms (tests that measure the electrical activity of the heart), and provoke symptoms such as stomach pain, diarrhoea, and vomiting. Long-term inhalation of antimony has been associated with pneumoconiosis (a lung disease caused by inhaling dust, leading to lung inflammation, and scarring) and hepatic (related to the liver) dysfunction, alongside blood abnormalities (Sunder and Chakravarty 2010). Notably, occupational exposure to antimony has been linked to an elevated risk of lung cancer, particularly among individuals with substantial occupational contact. Acute toxicity (harmful effects occurring shortly after exposure) resulting from antimony ingestion can manifest with severe symptoms like vomiting, abdominal cramps, and electrolyte imbalances (disturbances in essential minerals in the body). Concerns regarding antimony exposure in infants have been raised in relation to Sudden Infant Death Syndrome (SIDS)—the sudden, unexplained death of an otherwise healthy infant during sleep—although the precise mechanism remains elusive. Occupational studies suggest that antimony exposure may induce oxidative DNA damage (harm to DNA caused by reactive molecules) and immune system alterations, potentially affecting worker health. Furthermore, heightened levels of antimony in myocardial (related to the heart muscle) and muscular biopsy samples have been noted in individuals with idiopathic dilated cardiomyopathy (a condition where the heart's chambers are enlarged and weakened without a known cause), hinting at a potential association with cardiovascular health issues (Cooper and Harison 2009a, b).

10. **Barium (Ba)**

Barium (Ba) is primarily sourced from minerals like barite ($BaSO_4$) and witherite (barium carbonate). These minerals are commonly found in sedimentary rocks and are often associated with lead and zinc ores. Additionally, barium can be released into the environment through the weathering of rocks, industrial activities, and the use of barium-containing products, such as in the oil and gas industry and as a contrast agent in medical imaging. Inhalation exposure to barium may result in various adverse health effects, although available evidence is somewhat limited. Human case reports and animal studies indicate potential respiratory impacts, including benign pneumoconiosis (a lung disease caused by inhaling dust, leading to inflammation, and scarring) and silicosis (a lung disease caused by inhaling silica dust) observed in workers exposed to barium sulphate dust. Furthermore, both humans and animals exposed to barium have shown elevated blood pressure and cardiac irregularities (abnormal heart rhythms) (Essing et al. 1976; Hicks et al. 1986). Acute exposure to barium carbonate powder has been associated with gastrointestinal (related to the stomach and intestines) effects such as abdominal cramps and nausea (a feeling of sickness with an urge to vomit) in humans, while musculoskeletal (related to muscles and bones) effects like muscle weakness and paralysis (loss of the ability to move) have been noted in individuals inadvertently exposed to the same substance (Shankle and Keane 1988). Other reported effects encompass altered haematological (related to blood) parameters, impaired liver detoxification function (the liver's ability to remove toxins), renal (related to the kidneys) failure, and reduced body weight gain in animal studies (Tarasenko et al. 1977). Despite these findings indicating potential health risks linked to inhalation exposure to barium, further research is warranted to solidify definitive conclusions.

11. Beryllium (Be)

Beryllium (Be) is primarily sourced from geological formations, particularly from minerals like beryl ($Be_3Al_2Si_6O_{18}$) and chrysoberyl ($BeAl_2O_4$). These minerals are often found in igneous rocks, especially in pegmatites. Beryllium is also released into the environment through mining, industrial processes, and the use of beryllium-containing products. Occupational exposure to beryllium, commonly found in industries such as nuclear weapons production and aerospace, can provoke severe dermatitis (skin inflammation), reversible pneumonitis (lung inflammation that can improve with treatment), and a chronic granulomatous lung disease known as Chronic Beryllium Disease (CBD). The CBD, which is more prevalent in individuals with a genetic predisposition, differs from sarcoidosis (a disease involving abnormal collections of inflammatory cells in various organs) and necessitates specific immunological (related to the immune system) testing for accurate diagnosis. Workers in nuclear energy facilities face significant risks of beryllium exposure, which can lead to acute pneumonitis progressing to irreversible lung disease, particularly with copper-beryllium alloys. Beryllium oxide, utilized in ceramics, can also trigger delayed CBD effects. Primarily absorbed through the lungs, beryllium can induce respiratory symptoms such as coughing, chest pain, and dyspnoea (shortness of breath), as well as conjunctivitis (inflammation of the outer membrane of the eyeball and eyelid) and skin irritation. Chronic exposure heightens the risk of lung fibrosis (scarring of lung tissue) and cancer. Even when exposure remains below current safety thresholds, immunologically mediated hypersensitivity reactions characterized by granuloma (a small area of inflammation) formation may occur (Cooper and Harison 2009a, b).

12. Bismuth (Bi)

Bismuth (Bi) is primarily sourced from geological formations, typically found in minerals such as bismuthinite (Bi_2S_3) and bismite (Bi_2O_3). These minerals are usually located in hydrothermal veins and as a byproduct of mining operations for other metals. Although bismuth is generally considered non-toxic, prolonged use can lead to side effects and toxicity, particularly from the overuse of bismuth-containing medications. The degree of toxicity varies based on the type and quantity of bismuth absorbed, with soluble compounds posing higher risks for conditions such as neurotoxicity (damage to the nervous system) and nephrotoxicity (kidney damage). Bismuth tends to accumulate in organs like the kidneys, liver, and brain before being excreted in urine and faeces. Both acute (short-term) and chronic (long-term) exposures to bismuth can result in various toxic effects, including neurotoxicity, gastrointestinal toxicity (damage to the digestive system), nephrotoxicity, and hepatotoxicity (liver damage); however, these effects typically improve upon discontinuation of bismuth use. Bismuth Iodoform Paraffin Paste (BIPP), used in surgical procedures as an antiseptic, can occasionally induce severe adverse effects such as encephalopathy (a brain disorder characterized by confusion and an unsteady gait). The removal of BIPP generally leads to recovery, suggesting that bismuth may interfere with the metabolism of the central nervous system (Yang and Sun 2011).

13. **Cobalt (Co)**

Cobalt (Co) is an essential trace element found in vitamin B12, which is vital for red blood cell formation and neurological function. It primarily enters the human body through dietary sources, particularly animal products like meat and dairy. Geological sources of cobalt include minerals such as cobaltite (CoAsS) and erythrite ($Co_3(AsO_4)_2 \cdot 8H_2O$), typically mined from nickel and copper deposits. Industrial exposure to cobalt occurs in activities such as cobalt powder production, hard metal processing, and diamond polishing using cobalt disks, often combined with other substances like tungsten carbide and iron, which can influence cobalt's biological effects. Cobalt is primarily absorbed through the lungs and gastrointestinal tract, with minimal absorption through the skin, and is predominantly excreted through urine (Leyssens et al. 2017). Exposure to industrial cobalt poses several health risks, including skin and respiratory tract issues such as allergic dermatitis (a skin rash caused by an allergic reaction to substances), rhinitis (inflammation of the nasal mucous membrane, leading to symptoms like a runny nose and sneezing), and asthma (a chronic condition characterized by airway inflammation and difficulty breathing). Inhalation of cobalt-containing dust can lead to severe lung conditions known as hard metal disease, which is characterized by alveolitis (inflammation of the alveoli, the tiny air sacs in the lungs) and pulmonary fibrosis (scarring of lung tissue that can lead to breathing difficulties). Historical cases of cardiomyopathy (a disease of the heart muscle that affects its ability to pump blood) have been linked to the consumption of cobalt-fortified beer, raising concerns about excessive cobalt exposure. While there is no definitive evidence that cobalt alone increases the risk of lung cancer, combined exposure to cobalt and other substances, like tungsten carbide in the hard metal industry, may heighten this risk, necessitating further investigation (Lauwerys and Lison 1994). Cobalt toxicity can affect multiple organ systems; acute exposure may impact the endocrine system (the system that produces hormones regulating metabolism, growth, and reproduction), the cardiovascular system (the heart and blood vessels), the nervous system (the brain, spinal cord, and nerves), the gastrointestinal system (the digestive tract), and blood. Chronic inhalation of cobalt can lead to occupational asthma (asthma triggered by exposure to substances in the workplace) and hard metal disease. Cobalt ions can inhibit crucial enzymes (proteins that facilitate biochemical reactions), potentially resulting in cardiomyopathy, hypothyroidism (an underactive thyroid gland, which can lead to fatigue and weight gain), and disturbances in erythropoiesis (the process of producing red blood cells). Pulmonary toxicity (lung damage) arises from the generation of free radicals (unstable molecules that can damage cells and DNA), while dermatitis typically involves a type IV hypersensitivity reaction (an immune response that occurs after sensitization to a substance, resulting in delayed inflammation). Histopathologically, cobalt cardiomyopathy exhibits myocyte atrophy (wasting of heart muscle cells) and myofibril loss (loss of the contractile elements in muscle cells). Meanwhile, hard metal lung disease (HMLD) is characterized by multinucleated giant cells (large cells with multiple nuclei formed from the fusion of macrophages) and inflammatory infiltrates (accumulation of immune cells in tissues). *Arthroprosthetic cobaltism*, resulting from implant failure and systemic toxicity, involves lymphocytic perivascular infiltrates (infiltration of lymphocytes, a type of white blood cell, around blood vessels) and discoloration of synovial fluid (the fluid in joints that lubricates and nourishes cartilage) (Che and Lee 2023).

14. **Fluorine (F)**

Fluoride is mainly found in limestone and granites, as well as in minerals like fluorite (CaF_2) and cryolite (Na_3AlF_6). Natural processes like weathering and erosion release fluoride into the environment, which can contaminate groundwater and surface water. Human activities, including the production of phosphate fertilizers and industrial emissions, can also increase fluoride levels in soil and water. Fluoride can affect the human body in several ways, leading to various health issues. It is associated with metabolic and nutritional disorders, dental problems, and musculoskeletal conditions such as skeletal fluorosis, which is a bone disease caused by excessive fluoride intake, leading to pain and fractures. It may also impact reproductive and developmental health, cause neurotoxicity (toxic effects on the nervous system), and induce changes in behaviour. Additionally, fluoride exposure has been linked to genotoxicity (damage to genetic information), carcinogenicity (the potential to cause cancer), and disturbances in the gastrointestinal tract, kidneys, liver, and immune system (Guth et al. 2020). Health issues connected to fluoride exposure include skin problems like acne, arterial calcification (hardening of the arteries), and weakened bones that can result in fractures, including a specific type of bone cancer called osteosarcoma. Fluoride exposure may also increase risks for heart failure, cognitive deficits (problems with thinking and memory), and dental fluorosis, which causes discoloration and damage to teeth. Studies have shown links to diabetes, early onset of puberty in girls, and high blood pressure (hypertension). Other adverse effects may include impacts on the immune system, sleep disturbances, and potential contributions to iodine deficiency, which can affect thyroid function. Furthermore, fluoride exposure may reduce fertility rates, lower IQ, and cause damage to the heart muscle (myocardial damage). Neurotoxic effects include conditions like attention deficit hyperactivity disorder (ADHD) and disorders affecting the jaw joint (temporomandibular joint disorder, TMJ) (Kall and Cole 2024).

15. **Silver (Ag)**

Silver is primarily sourced from geological deposits, often in the form of native silver or as part of silver sulphides like argentite (Ag_2S). It is commonly extracted from ores that contain lead, copper, or zinc, and can also be found in epithermal and placer deposits. These geological sources are vital for silver's production in various industrial applications. Chronic exposure to silver can result in two notable adverse effects: argyria, characterized by a permanent bluish-grey discoloration of the skin or eyes, and argyrosis. These effects are primarily linked to exposure to soluble forms of silver. Alongside discoloration, soluble silver compounds can induce liver and kidney damage, irritate the eyes, skin, respiratory, and intestinal tract, and lead to alterations in blood cells. However, metallic silver appears to pose minimal health risks. Current occupational exposure limits do not differentiate between soluble and metallic silver, prompting recommendations for separate permissible exposure limits (PELs) by researchers (Drake and Hazelwood 2005).

16. **Lithium (Li)**

Lithium is mainly obtained from geological deposits like pegmatites and brines, where it is found in minerals such as spodumene (LiAl $(SiO_3)_2$) and lepidolite $(KLi_2Al(Si_4O_{10})(F,OH)_2)$. The extraction methods usually involve open-pit mining and evaporating brine, making geological knowledge important for effective lithium recovery. When it comes to lithium toxicity, there are various neurological effects, including coarse tremors (shaking), hyperreflexia (overactive reflexes), nystagmus (involuntary eye movements), and ataxia (difficulty coordinating movements). These symptoms can affect a person's consciousness, ranging from mild confusion to delirium, and while they are usually reversible, some may last up to 12 months. Renal (kidney) toxicity is more common in people who take lithium long-term, leading to issues such as trouble concentrating urine, nephrogenic diabetes insipidus (where kidneys can't concentrate urine), sodium-losing nephritis (kidney inflammation), and nephrotic syndrome (excess protein in urine). Cardiovascular (heart) effects, though often mild, can include T wave flattening (Flat T waves are one of the abnormalities associated with T waves on an ECG) and heart rhythm issues, which are reversible. Gastrointestinal (stomach and digestive system) symptoms often appear within an hour of overdose. Endocrine (hormonal) effects include reduced production of thyroid hormones, causing hypothyroidism (low thyroid hormone levels). In rare cases, hyperthyroidism (high thyroid hormone levels) can worsen lithium toxicity by affecting how the kidneys handle lithium (Hedya et al. 2023). Lithium is commonly used to treat bipolar disorder, but it can lead to several health risks. Sexual dysfunction, such as decreased sexual desire and satisfaction, is a common side effect, especially in bipolar patients. Skin issues like acne and psoriasis may also arise, requiring dose changes or alternative medications. Lithium intoxication, which includes neurological symptoms like ataxia and confusion, needs careful monitoring and immediate action, such as stopping the medication or using dialysis in severe cases (Gitlin 2016). Long-term use can cause serious damage to organs. The kidneys might suffer irreversible harm, shown by kidney shrinkage and scarring, which can result in excessive urination and kidney failure (Ferensztajn-Rochowiak and Rybakowski 2023). Thyroid issues are common, ranging from low thyroid function to goitre (enlarged thyroid). Lithium is also linked to high calcium levels in the blood and overactive parathyroid glands, though these symptoms are usually milder than those of primary hyperparathyroidism. While lithium is effective for managing bipolar disorder, careful monitoring of potential health problems is crucial for patient safety. Regular checks on sexual function, skin issues, kidney function, thyroid hormone levels, and calcium levels are essential parts of lithium therapy management. Close cooperation between patients, doctors, and specialists is important to reduce risks and improve treatment outcomes (Gitlin 2016).

17. Palladium (Pd)

Palladium (Pd) is primarily sourced from geological deposits associated with nickel and copper mining, particularly in regions like the Bushveld Igneous Complex in South Africa, the Norilsk-Talnakh region in Russia, and the Stillwater Complex in Montana, USA. It is mainly found in sulfide minerals such as pentlandite $[(Ni,Fe)_9S_8]$ and chalcopyrite $(CuFeS_2)$. During the refining of these ores, palladium is extracted as a byproduct. Additionally, it can be recovered from recycled materials, particularly automotive catalytic converters, where it reduces harmful emissions. Palladium poses health concerns due to its sensitization risk, especially for individuals with nickel allergies, impacting workers in industries like mining, dental technology, and chemicals. Pd exposure can lead to skin and eye irritations, while the general population may encounter it through dental restorations and jewellery. Protective measures, such as using corrosion-resistant alloys, are essential to minimize Pd release. Individuals sensitive to Pd should avoid Pd-containing materials, although some tolerate them without adverse effects. Palladium nanoparticles (Pd-NPs) can induce varying cytotoxic effects, including apoptosis and DNA damage, influenced by their surface coating and chemical composition. Palladium nanoparticles' exposure raises concerns about oxidative stress and immune modulation, with IL-8 secretion serving as a biomarker for inflammation. Further research is necessary to understand their long-term health effects (Kielhorn et al. 2002; Leso and Iavicoli 2018).

18. Vanadium (V)

Vanadium exposure poses significant respiratory risks for both humans and animals. In humans, exposure to vanadium, particularly vanadium pentoxide, can result in persistent coughing, wheezing, chest pain, and respiratory distress. Even at low exposure levels, individuals may experience symptoms such as coughing. Animal studies corroborate these findings, showing signs of respiratory distress and impaired lung function in response to vanadium exposure. Additionally, vanadium exposure has been associated with cardiovascular, gastrointestinal, haematological, musculoskeletal, hepatic, renal, dermal, ocular, immunological, and neurological effects, albeit to varying extents. Workers exposed to vanadium dusts have reported weight loss and occasional neurological symptoms such as dizziness or tremors (Toxicological Profile for Vanadium 2012).

19. Zinc (Zn)

Zinc, an essential trace element, presents diverse exposure pathways, each carrying distinct health implications. Zinc is primarily sourced from minerals like sphalerite

(ZnS), often found alongside lead, silver, and copper ores. Environmental pathways for zinc exposure stem from mining, smelting, and industrial activities, which release zinc into air, water, and soil. Inhalation of zinc-containing smoke, prevalent in industrial settings and military operations, can lead to respiratory issues such as metal fume fever (MFF), while dermal exposure, though less studied, may cause irritation, particularly with zinc chloride. Orally, excessive zinc intake can induce gastrointestinal symptoms and potentially fatal copper deficiency, affecting immune function and even increasing cancer risk. Additionally, zinc plays a role in apoptosis (programmed cell death) and neuronal damage, highlighting its impact beyond toxicity. Both deficiency and excess can yield significant health issues, making the balance of adequate zinc intake crucial (Plum et al. 2010).

20. **Copper (Cu)**

Chronic copper toxicity, or an excess of copper in the body over a long period, primarily affects the liver. This is observed in individuals with Wilson disease, a genetic disorder where the body cannot properly eliminate copper, leading to its buildup in the liver and brain, and in children with certain cirrhosis syndromes, conditions where the liver slowly deteriorates. Copper naturally occurs in minerals such as chalcopyrite ($CuFeS_2$), bornite (Cu_5FeS_4), and malachite ($Cu_2CO_3(OH)_2$). These minerals are significant sources of copper and enter the environment through mining, smelting (the process of extracting metals from ores), and contamination of water systems. In rare cases, prolonged intake of high-dose copper supplements can lead to liver diseases. Neurological problems, such as tremors or cognitive issues, are commonly observed in individuals with Wilson disease or animals exposed to very high levels of copper. Another effect of elevated copper levels is haemolytic anaemia, where red blood cells are destroyed faster than the body can produce them, particularly in instances of acute hepatic necrosis, or sudden, severe liver damage. The effects of excess copper on human reproduction and development are not well-studied, and evidence from animal studies is mixed. Research on the connection between copper exposure and cancer has produced inconsistent results, without conclusive evidence of a direct link. Copper has potential genotoxic (DNA-damaging) and mutagenic (mutation-causing) effects, but studies offer conflicting conclusions. Certain populations, such as infants, young children, and individuals with genetic conditions like Wilson disease, may be more sensitive to copper exposure. Animal studies offer insights into copper's toxicological effects, but due to differences in copper intake levels and exposure pathways, the results may not fully apply to humans (National Research Council 2000a, b) (Fig. 3.3).

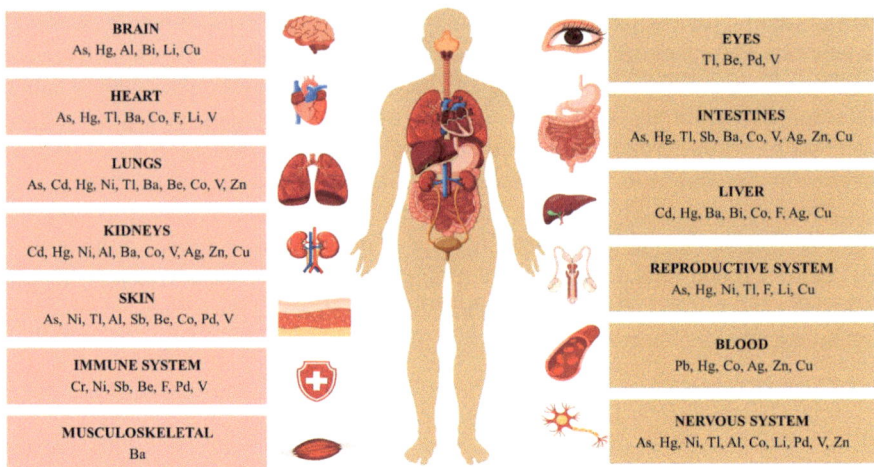

Fig. 3.3   Sketch Showing organs affected by specific trace elements

References

Antoniadis V, Shaheen SM, Levizou E, Shahid M, Niazi NK, Vithanage M, Ok YS, Bolan N, Rinklebe J (2019) A critical prospective analysis of the potential toxicity of trace element regulation limits in soils worldwide: are they protective concerning health risk assessment? a review. Environ Int 127:819–847. ISSN 0160-4120. https://doi.org/10.1016/j.envint.2019.03.039

Alsanosi SMM, Skiffington C, Padmanabhan S (2014) Pharmacokinetic pharmacogenomics. In: Handbook of pharmacogenomics and stratified medicine. https://doi.org/10.1016/b978-0-12-386882-4.00017-7

Alekseenko SI, Skalny AV, Ajsuvakova OP, Skalnaya MG, Notova SV, Tinkov AA (2024) Trace elements and metals—LiverTox—NCBI bookshelf. nih.gov. https://www.ncbi.nlm.nih.gov/books/NBK548854/

Balali-Mood M, Naseri K, Tahergorabi Z, Khazdair MR, Sadeghi M (2021) Toxic mechanisms of five heavy metals: mercury, lead, chromium, cadmium, and arsenic. Front Pharmacol. https://doi.org/10.3389/fphar.2021.643972

Barnes JL, Zubair M, John K, Poirier MC, Martin FL (2018) Carcinogens and DNA damage. Biochem Soc Trans. https://doi.org/10.1042/bst20180519

Bellinger D, Leviton A, Waternaux C, Needleman H, Rabinowitz M (1987) Longitudinal analyses of prenatal and postnatal lead exposure and early cognitive development. N Engl J Med 316:1037–1043

Bellinger DC (2005) Teratogen update: lead and pregnancy. Birth Defects Res A Clin Mol Teratol 73:409–420

Behrens T, Ge C, Vermeulen R, Kendzia B, Olsson A, Schuz J, Kromhout H, Pesch B, Peters S, Portengen L et al (2023) Occupational exposure to nickel and hexavalent chromium and the risk of lung cancer in a pooled analysis of case-control studies (SYNERGY). Int J Cancer 152:645–660

Blake J, Rosenblum ND (2014) Renal branching morphogenesis: morphogenetic and signaling mechanisms. Semin Cell Dev Biol. https://doi.org/10.1016/j.semcdb.2014.07.011

Brunton LL, Goodman LS, Blumenthal D, Buxton I, Parker KL (eds) (2007) Goodman and Gillmans manual of pharmocology and theraupetics. Mcgraw Hill Professional

Cavallazzi R, Folz RJ (2012) Pulmonary complications of hematopoietic stem cell transplantation. Clin Respir Med. https://doi.org/10.1016/b978-1-4557-0792-8.00077-5

Chen RJ, Lee VR (2024) Cobalt toxicity. [Updated 2023 Jul 29]. In: StatPearls [Internet]. Treasure Island (FL): StatPearls Publishing. Available from: https://www.ncbi.nlm.nih.gov/books/NBK 587403/

Chisolm J, Harrison H (1956) The exposure of children to lead. J Am Acad Pediatrics. 18:943–958

Charkiewicz AE, Omeljaniuk WJ, Nowak K, Garley M, Nikliński J (2023) Cadmium toxicity and health effects—a brief summary. Molecules 28(18):6620. https://doi.org/10.3390/molecules281 86620. PMID: 37764397; PMCID: PMC10537762

Cooper RG, Harrison AP (2009a) The exposure to and health effects of antimony. Ind J Occup Environ Med. 13(1):3–10. https://doi.org/10.4103/0019-5278.50716. PMID: 20165605; PMCID: PMC2822166

Cooper RG, Harrison AP (2009b) The uses and adverse effects of beryllium on health. Ind J Occup Environ Med 13(2):65–76. https://doi.org/10.4103/0019-5278.55122. PMID: 20386622; PMCID: PMC2847329

Dart RC, Hurlbut KM, Boyer-Hassen LV (2004) Lead. In: Dart RC (ed) Medical toxicology, 3rd ed. Lippincot Williams and Wilkins

Das KK, Das SN, Dhundasi SA (2008) Nickel, its adverse health effects & oxidative stress. Ind J Med Res 128(4):412–425. PMID: 19106437

Davis PJ, Bosenberg A, Davidson A, Jimenez N, Kharasch E, Lynn AM, Tofovic SP, Woelfel S (2011) Pharmacology of pediatric anesthesia. Smith's anesthesia for infants and children. https://doi.org/10.1016/b978-0-323-06612-9.00007-9

Drake PL, Hazelwood KJ (2005) Exposure-related health effects of silver and silver compounds: a review. Ann Occup Hyg 49(7):575–585. https://doi.org/10.1093/annhyg/mei019. Epub 2005 Jun 17. PMID: 15964881

Division of Research Safety | Illinois (2024) illinois.edu. https://drs.illinois.edu/Page/SafetyLib rary/HealthEffectsOfChemicalExposure

Deng R, Gao J, Yi J, Liu P (2022) Could peony seeds oil become a high-quality edible vegetable oil? The nutritional and phytochemistry profiles, extraction, health benefits, safety and value-added-products. Food Res Int. https://doi.org/10.1016/j.foodres.2022.111200

Endocrine Disruptors (2024) National Institute of Environmental Health Sciences. https://www.niehs.nih.gov/health/topics/agents/endocrine

Essing H-G, Buhlmeyer G, Valentin H et al (1976) [Exclusion of disturbances to health from long years of exposure to barium carbonate in the production of steatite ceramics.] Arbeitsmedizin Sozialmedizin Praventivmedizin 11:299–302 (German)

Farzan SF, Karagas MR (2013) In utero and early life arsenic exposure in relation to long-term health and disease. Toxicol Appl Pharmacol; Chen Y (2013) 272(2):384–390

Ferensztajn-Rochowiak E, Rybakowski JK (2023) Long-term lithium therapy: side effects and interactions. Pharmaceuticals (Basel) 16(1):74. https://doi.org/10.3390/ph16010074. PMID: 36678571; PMCID: PMC9867198

Genchi G, Carocci A, Lauria G, Sinicropi MS, Catalano A (2020a) Nickel: human health and environmental toxicology. Int J Environ Res Public Health 17(3):679. https://doi.org/10.3390/ijerph17030679. PMID: 31973020; PMCID: PMC7037090

Genchi G, Sinicropi MS, Lauria G, Carocci A, Catalano A (2020b) The effects of cadmium toxicity. Int J Environ Res Pub Health 17(11):3782. https://doi.org/10.3390/ijerph17113782. PMID: 32466586; PMCID: PMC7312803

Genchi G, Carocci A, Lauria G, Sinicropi MS, Catalano A (2021) Thallium use, toxicity, and detoxification therapy: an overview. Appl Sci 11(18):8322. https://doi.org/10.3390/app111 88322

Geyer HJ, Schheunert I, Rapp K, Gebefugi I, Steinberg C, Kettrup A (1993) The relevance of fat content in toxicity of lipophilic chemicals to terrestrial animals with special reference to dieldrin and 2,3,7,8-tetrachlorodibenzo-p-dioxin (TCDD). Ecotoxicol Environ Saf. https://doi.org/10.1006/eesa.1993.1040. PMID: 32466586; PMCID: PMC7312803

Goering PL, Barber DS (2010) Hepatotoxicity of copper, iron, cadmium, and arsenic. Compr Toxicol. https://doi.org/10.1016/b978-0-08-046884-6.01022-8

Gooch J (2023) Basics of dose-response—toxicology education foundation. Toxicology education foundation—TEF provides the public with access to reliable toxicology scientific information in order to make informed decisions about chemical exposures. https://toxedfoundation.org/basics-of-dose-response/

Gregus Z, Klaassen CD (2001) Mechanisms of toxicity. In: Klaassen CD (ed) Casarett and Doull's toxicology: the basic science of poisons, 6th edn. McGraw-Hill, New York, pp 35–81

Gregus Z (1986) Disposition of metals in rats: a comparative study of fecal, urinary, and biliary excretion and tissue distribution of eighteen metals. Toxicol Appl Pharmacol. https://doi.org/10.1016/0041-008x(86)90384-4

Guth S, Hüser S, Roth A, Degen G, Diel P, Edlund K, Eisenbrand G, Engel KH, Epe B, Grune T, Heinz V, Henle T, Humpf HU, Jäger H, Joost HG, Kulling SE, Lampen A, Mally A, Marchan R, Marko D, Mühle E, Nitsche MA, Röhrdanz E, Stadler R, van Thriel C, Vieths S, Vogel RF, Wascher E, Watzl C, Nöthlings U, Hengstler JG (2020) Toxicity of fluoride: critical evaluation of evidence for human developmental neurotoxicity in epidemiological studies, animal experiments and in vitro analyses. Arch Toxicol 94(5):1375–1415. https://doi.org/10.1007/s00204-020-02725-2. Epub 2020 May 8. PMID: 32382957; PMCID: PMC7261729

Gitlin M (2016) Lithium side effects and toxicity: prevalence and management strategies. Int J Bipolar Disord 4(1):27. https://doi.org/10.1186/s40345-016-0068-y. Epub 2016 Dec 17. PMID: 27900734; PMCID: PMC5164879

Hagvall L, Pour MD, Feng J, Karma M, Hedberg Y, Malmberg P (2021) Skin permeation of nickel, cobalt and chromium salts in ex vivo human skin, visualized using mass spectrometry imaging. Toxicol In Vitro 76:105232

Henretig FM (2006) Lead. In: Golgfrank LR (ed) Goldfrank's toxicologic emergencies, 8th ed. McGraw Hill Professional

Hedya SA, Avula A, Swoboda HD (2024) Lithium toxicity. [Updated 2023 Jun 26]. In: StatPearls [Internet]. Treasure Island (FL): StatPearls Publishing. Available from: https://www.ncbi.nlm.nih.gov/books/NBK499992/

Health Effects of Exposure to Substances and Carcinogens (2024a) Toxic substance portal | ATSDR. cdc.gov. https://wwwn.cdc.gov/TSP/substances/ToxOrganSystems.aspx

Hicks R, Caldas LQ A, Dare PR M et al (1986) Cardiotoxic and bronchoconstrictor effects of industrial metal fumes containing barium. Arch Toxicol Suppl 9 Toxic interfaces of neurones, smoke and genes. Springer-Verlag New York, Inc., Secaucus, NJ

introtoxsubstances.pdf. (2017). ca.gov. https://www.cdph.ca.gov/Programs/CCDPHP/DEODC/OHB/HESIS/CDPH%20Document%20Library/introtoxsubstances.pdf

Jackson E, Shoemaker R, Larian N, Cassis L (2017) Adipose tissue as a site of toxin accumulation. Compr Physiol. https://doi.org/10.1002/cphy.c160038

Jia N, Guo C, Nakazawa Y, van den Heuvel D, Luijsterburg MS, Ogi T (2021) Dealing with transcription-blocking DNA damage: repair mechanisms, RNA polymerase II processing and human disorders. DNA Repair. https://doi.org/10.1016/j.dnarep.2021.103192

Kall and Cole (2024) Fluoride exposure and human health risk. https://iaomt.org/resources/fluoride-facts/fluoride-exposure-human-health-risks/

Kemnic TR, Coleman M (2023) Thallium toxicity. [Updated 2023 Jul 10]. In: StatPearls [Internet]. Treasure Island (FL): StatPearls Publishing. Available from: https://www.ncbi.nlm.nih.gov/books/NBK513240/

Kielhorn J, Melber C, Keller D, Mangelsdorf I (2002) Palladium—a review of exposure and effects to human health. Int J Hyg Environ Health. 205(6):417–432. https://doi.org/10.1078/1438-4639-00180. PMID: 12455264

Klotz K, Weistenhöfer W, Neff F, Hartwig A, van Thriel C, Drexler H (2017) The health effects of aluminium exposure. Dtsch Arztebl Int 114(39):653–659. https://doi.org/10.3238/arztebl.2017.0653. PMID: 29034866; PMCID: PMC5651828

Kosnett MJ (2005) Lead. In: Brent J (ed) Critical care toxicology: diagnosis and management of the critically poisoned patient. Gulf Professional Publishing. ISBN 0-8151-4387-7

Lauwerys R, Lison D (1994) Health risks associated with cobalt exposure—an overview. Sci Total Environ 150(1–3):1–6. ISSN 0048-9697 https://doi.org/10.1016/0048-9697(94)90125-2

Langman LJ, Kapur BM (2006) Toxicology: then and now. Clin Biochem 39:498–510

Landrigan PJ, Schechter CB, Lipton JM, Fahs MC, Schwartz J (2002) Environmental pollutants and disease in American children. Environ Health Perspect 110:721–728

Leyssens L, Vinck B, Van Der Straeten C, Wuyts F, Maes L (2017) Cobalt toxicity in humans—a review of the potential sources and systemic health effects. Toxicology 387:43–56. ISSN 0300-483X, https://doi.org/10.1016/j.tox.2017.05.015

Leso V, Iavicoli I (2018) Palladium nanoparticles: toxicological effects and potential implications for occupational risk assessment. Int J Mol Sci 19(2):503. https://doi.org/10.3390/ijms19020503. PMID: 29414923; PMCID: PMC5855725

Lehman-McKeeman LD (2024) Absorption, distribution, and excretion of toxicants. McGraw Hill Medical. https://accesspharmacy.mhmedical.com/content.aspx?bookid=1540&sectionid=92525314

Liu G, Wang J, Liu X, Liu X, Li X, Ren Y, Wang J, Dong L (2018) Partitioning and geochemical fractions of heavy metals from geogenic and anthropogenic sources in various soil particle size fractions. Geoderma 312:104–113

Mahaffey KR (2005) Mercury exposure: medical and public health issues. Trans Am Clin Climatol Assoc 116:127–153; discussion 153–154. PMID: 16555611; PMCID: PMC1473138

Mandal GC, Mandal A, Chakraborty A (2023) The toxic effect of lead on human health: a review. Hum Biol Pub Health 3. https://doi.org/10.52905/hbph2022.3.45

Merill JC, Morton JJP, Soileau SD. Metals. In: Hayes AW (ed) Principles and methods of toxicology, 5th ed. CRC Press

Mycyk M, Hryhorcu D, Amitai Y (2005) Lead. In: Erickson TB, Ahrens WR, Aks S, Ling L (eds) Paediatric toxicology: diagnostic and management of the poisoned child. Mcgraw Hill Professional

National Research Council (US) (1989) Committee on diet and health. Diet and health: implications for reducing chronic disease risk. National Academies Press (US), Washington (DC), Trace elements. Available from: https://www.ncbi.nlm.nih.gov/books/NBK218751/

National Research Council (US) (2000a) Commission on engineering and technical systems. National Research Council (US) Commission on life sciences; McKone TE, Huey BM, Downing E et al (eds) Strategies to protect the health of deployed U.S. forces: detecting, characterizing, and documenting exposures. National Academies Press (US), Washington (DC). Environmental and exposure pathways. Available from: https://www.ncbi.nlm.nih.gov/books/NBK225345/

National Research Council (US) (2000b) Committee on copper in drinking water. Copper in drinking water. National Academies Press (US), Washington (DC). Health effects of excess copper. Available from: https://www.ncbi.nlm.nih.gov/books/NBK225400/

Patrick L (2006) Lead toxicity, a review of the literature. Part 1: Exposure, evaluation, and treatment. Altern Med Rev 11:2–22

Pearce JMS (2007) Burton's line in lead poisoning. Eur Neurol 57:118–119

Plum LM, Rink L, Haase H (2010) The essential toxin: impact of zinc on human health. Int J Environ Res Public Health 7(4):1342–1365. https://doi.org/10.3390/ijerph7041342. Epub 2010 Mar 26. PMID: 20617034; PMCID: PMC2872358

Plasma Protein Binding (2021) pharmainformatic.com. https://www.pharmainformatic.com/html/plasma_protein_binding.html

Pourret O, Bollinger J-C (2018) Heavy metal-what to do now: to use or not to use. Sci Total Environ 610–611:419–420

Pourentezari M (2024) Effects of acrylamide on sperm parameters, chromatin quality, and the level of blood testosterone in mice. PubMed Central (PMC). https://www.ncbi.nlm.nih.gov/pmc/articles/PMC4094659/

Primak L, Blumer JL (2006) Principles of drug disposition in the critically ill child. Pediatr Crit Care. https://doi.org/10.1016/b978-032301808-1.50113-9

Quansah R, Armah FA, Essumang DK, Luginaah I, Clarke E, Marfoh K et al (2015) Association of arsenic with adverse pregnancy outcomes/infant mortality: a systematic review and meta-analysis. Environ Health Perspect 123(5):412–21

Randive K (2013) Elements of geochemistry, geochemical exploration and medical geology

Radfard M, Hashemi H, Baghapour MA, Samaei MR, Yunesian M, Soleimani H, Azhdarpoor A (2023) Prediction of human health risk and disability-adjusted life years induced by heavy metals exposure through drinking water in Fars Province. Scientific Reports, Iran. https://doi.org/10.1038/s41598-023-46262-1

Rinklebe J, Antoniadis V, Shaheen SM, Rosche O, Altermann M (2019) Health risk assessment of potentially toxic elements in soils along the Central Elbe River, Germany. Environ Int 126:76–88. https://doi.org/10.1016/j.envint.2019.02.011

Schmid RD (2024) Absorption, distribution, metabolism, and excretion of toxic agents in animals—absorption, distribution, metabolism, and excretion of toxic agents in animals—Merck Veterinary Manual. Merck Veterinary Manual. https://www.merckvetmanual.com/toxicology/toxicology-introduction/absorption-distribution-metabolism-and-excretion-of-toxic-agents-in-animals

Shankle R, Keane JR (1988) Acute paralysis from inhaled barium carbonate. Arch Neurol 45:579–580

Shannon M (2003) Severe lead poisoning in pregnancy. AmbulPediatr 3:37–39

Shanker A (2008) Mode of action and toxicity of trace elements. https://doi.org/10.1002/9780470370124.ch21

Share (2024) Environmental mutagens and gene expression I learn science at scitable. nature.com. http://www.nature.com/scitable/topicpage/environmental-mutagens-cell-signalling-and-dna-repair-1090

Share (2024) Soil minerals and plant nutrition I learn science at scitable. nature.com. https://www.nature.com/scitable/knowledge/library/soil-minerals-and-plant-nutrition-127881474/?error=cookies_not_supported&code=50474fc4-5247-4391-9f89-eb247f6cae32

Sharma P, Singh SP, Parakh SK, Tong YW (2022) Health hazards of hexavalent chromium (Cr (VI)) and its microbial reduction. Bioengineered 13(3):4923–4938. https://doi.org/10.1080/21655979.2022.2037273. PMID: 35164635; PMCID: PMC8973695

Slitt AL (2024) Absorption, distribution, and excretion of toxicants. McGraw Hill Medical. https://accesspharmacy.mhmedical.com/content.aspx?bookid=2462&sectionid=194919337

Stoppler M (2024) Definition of toxicity

Sullivan PJ, Agardy FJ, Clark JJJ (2005) Living with the risk of polluted water. Environ Sci Drink Water. https://doi.org/10.1016/b978-075067876-6/50007-5

Sundar S, Chakravarty J (2010) Antimony toxicity. Int J Environ Res Public Health 7(12):4267–4277. https://doi.org/10.3390/ijerph7124267. Epub 2010 Dec 20. PMID: 21318007; PMCID: PMC3037053

Taking an Exposure History (2024b) Which organ systems are affected by toxic exposure(s) I environmental medicine I ATSDR. cdc.gov. https://www.atsdr.cdc.gov/csem/exposure-history/Organ-Systems-Are-Affected.html

Tarasenko NY, Pronin OA, Silaev AA (1977) Barium compounds as industrial poisons (an experimental study). J Hyg Epidemiol Microbiol Immunol 21:361–373

Tchounwou PB, Yedjou CG, Patlolla AK, Sutton DJ (2012) Heavy metal toxicity and the environment. Exp Suppl 101:133–64. https://doi.org/10.1007/978-3-7643-8340-4_6.PMID:22945569; PMCID:PMC4144270

Timbrell JA (ed) (2008) Principles of biochemical toxicology, 4th ed. Informa Health Care. Biochemical mechanisms of toxicity: specific examples

Toxicological Profile for Vanadium. Atlanta (GA) (2012) Agency for toxic substances and disease registry (US). Health effects. Available from: https://www.ncbi.nlm.nih.gov/books/NBK592340/

Tolins M, Ruchirawat M, Ann LP (2014) The developmental neurotoxicity of arsenic: cognitive and behavioural consequences of early life exposure. Glob Health 80(4):303–314

Toxicity (2024) cornell.edu. https://ehs.cornell.edu/book/export/html/1388

Welcome to ToxTutor (2024) Toxicology MSDT. https://www.toxmsdt.com/34-organ-specific-toxic-effects.html

Wu W, Wu P, Yang F, Sun DL, Zhang DX, Zhou YK (2018) Assessment of heavy metal pollution and human health risks in urban soils around an electronics manufacturing facility. Sci Total Environ 630:53–61

What You Know Can Help You—an introduction to toxic substances (2024). ny.gov. https://www.health.ny.gov/environmental/chemicals/toxic_substances.htm

Yang N, Sun H (2011) Bismuth: environmental pollution and health effects. Encycl Environ Health 414–420. https://doi.org/10.1016/B978-0-444-52272-6.00374-3. Epub 2011 Mar 3. PMCID: PMC7151860

Yu Y-Q, Yang X, Wu X-F, Fan Y-B (2021) Enhancing permeation of drug molecules across the skin via delivery in nanocarriers: novel strategies for effective transdermal applications. Front Bioeng Biotechnol. https://doi.org/10.3389/fbioe.2021.646554

Zargari F, Rahaman MS, KazemPour R, Hajirostamlou M (2022) Arsenic, oxidative stress and reproductive system. J Xenobiot 12(3):214–222. https://doi.org/10.3390/jox12030016. PMID: 35893266; PMCID: PMC9326564

Zulaikhah S, Wahyuwibowo J, Pratama A (2020) Mercury and its effect on human health: a review of the literature. Int J Pub Health Sci (IJPHS) 9:103–114. https://doi.org/10.11591/ijphs.v9i2.20416

Chapter 4
Geo-Diseases: An Introduction

This chapter introduces geo-diseases, which are illnesses arising from the Earth's natural processes and human interactions with the geological environment. These diseases may result from natural geohazards like earthquakes, volcanic eruptions, and landslides; or from direct exposure to toxic elements through activities like mining, where elements previously locked in rocks and minerals become more accessible. Furthermore, environmental disruptions and human activities can indirectly contribute to disease spread. The chapter is organized to first explore diseases linked to direct exposure pathways, followed by a discussion of occupational health risks associated with mining. It then examines illnesses triggered by geohazards and concludes with other related diseases.

4.1 What Are Geo-Diseases?

Geo-Diseases are health conditions resulting from exposure to naturally occurring geological materials, such as mineral dust, toxic gases, or heavy metals originating from the Earth. These materials can enter the human body through inhalation, ingestion, or skin absorption, leading to a range of illnesses. In addition to direct exposure, natural events such as earthquakes, volcanic eruptions, and landslides can indirectly contribute by contaminating air, water, and soil, creating conditions conducive to disease outbreaks. For example, volcanic eruptions release ash and sulphur dioxide, which can impair respiratory health, while landslides may disturb water bodies, resulting in waterborne diseases (Mavrouli et al. 2023; Mueller et al. 2020). Anthropogenic activities also play a significant role in the outbreak of geo-diseases. Humans themselves are also carriers of various pathogens which interact with the body as an outcome of environmental fluctuations. Occupational exposure of toxic elements and other geomaterials such as coal dust, to the people working in mining and construction industries is also an important cause of geodiseases. Chronic exposure to silica

K. Randive and P. Godbole, *Medical Geology for Beginners*,
SpringerBriefs in Medical Earth Sciences, https://doi.org/10.1007/978-3-031-82765-5_4

dust can cause silicosis, while mercury used in small-scale gold mining is associated with neurological and renal damage (Requena-Mullor et al. 2021; Cicek-Senturk et al. 2014). Furthermore, environmental degradation catalyses the spread of Geo-Diseases by jeopardizing the ecological systems. The impacts of climate change, including increased dust storms and vector-borne diseases, further amplify these risks (Hasan 2021). Thus, *"the geological diseases or the Geo-Diseases are those that are caused due to geohazards such as earthquake, volcanism, landslide, etc.; or direct exposure to toxic elements due to mining activity that are otherwise sequestered into geological reservoirs such as rocks and minerals; or indirectly through environmental disturbances and human activities"*.

4.2 Classification of Geogenic Diseases

Geogenic diseases include a variety of illnesses that are influenced by geological factors. These diseases can be caused by direct contact with the geological materials as well as by natural events (geohazards) or human (anthropogenic) activities. To better understand geogenic diseases, we can classify them into three main categories:

1. **Illnesses Caused by Direct Interaction with the Geological Materials**

   These diseases occur when people come into direct contact with substances from the Earth, such as minerals, dust, or gases. For example, inhaling silica dust can lead to silicosis, while exposure to asbestos can cause asbestosis.

2. **Illnesses Caused by Geohazards and Natural Agencies**

   Some diseases arise because of natural events like earthquake, volcanic eruption, and landslide. These events can pollute water, air, and soil, creating conditions that lead to illnesses. For example, volcanic ash can cause respiratory problems, and earthquakes can contaminate water, leading to waterborne diseases.

3. **Illnesses Caused by Humans in Response to Environmental Changes**

   Human activities, such as mining, can also cause diseases. These activities disturb the natural environment, releasing harmful substances into the air, water, or soil.
      In simple terms, this classification can be viewed as direct and indirect forms of diseases:

Direct diseases occurring due to direct exposure to toxic chemicals released due to mining activity or geological materials such as silica dust, coal dust, asbestos, etc.

Indirect diseases resulting from the aftermaths of the geohazards, such as outbreak of vector borne diseases. Or due to anthropogenic activities that affect the environment, such as, excessive withdrawal of water due to mining activity, contamination (pollution) of air, water, and soil. The ergonomic diseases caused due to occupational activities, such as fire, roof collapse, noise, postural injuries, etc. Similarly, chemical, radioactive, and electrical diseases; as well as psychological diseases to miners are among the indirect diseases.

4.3 Geo-Diseases Arising Due to Direct Exposure

4.3.1 Silicosis and Other Related Diseases

Exposure to respirable crystalline silica (SiO_2) poses significant health risks to workers, leading to well-known disease called silicosis (a lung condition caused by inhaling fine silica particles). These particles settle deep in the lungs, causing scarring that makes breathing difficult. This damage is permanent and worsens with continued exposure (Requena-Mullor et al. 2021). Workers exposed to silica dust are also at a higher risk of cardiovascular diseases (heart and blood vessel problems) and pulmonary tuberculosis (a bacterial infection of the lungs), as long-term silica exposure weakens the immune system in the lungs, making infections more likely to happen (Requena-Mullor et al. 2021). Chronic obstructive pulmonary disease (COPD) including chronic bronchitis (long-term inflammation of the airways) and emphysema (damage to the air sacs in the lungs), is another significant risk, leading to severe breathing difficulties. One of the most serious concerns is the lung cancer, as studies show that long-term inhalation of silica dust increases the risk of lung cancer. Silica exposure is also linked to cancers of the oesophagus (the tube that carries food to the stomach), stomach, and skin. This happens because silica particles produce reactive oxygen species (ROS), harmful chemicals that damage lung tissue and can lead to cancer. In 1997, the International Agency for Research on Cancer (IARC) classified crystalline silica as a Group 1 carcinogen, confirming its ability to cause cancer in humans (Pelucchi et al. 2006). Silica exposure has also been linked to autoimmune diseases (where the immune system mistakenly attacks the body's own tissues) and to kidney disorders (which affect the kidneys' ability to filter waste from the blood) (Requena-Mullor et al. 2021). To reduce these risks, industries such as mining, stone cutting, and construction must use protective gear, apply dust control measures, and follow safety regulations to safeguard workers' health (Requena-Mullor et al. 2021).

4.3.2 Coal Workers' Pneumoconiosis (CWP) and Associated Diseases

Health problems in coal production arise from dust exposure and the body's response, particularly pneumoconiosis (lung disease caused by inhaling dust). Coal workers' pneumoconiosis (CWP), which is also known as *black lung*, refers to various conditions resulting from long-term exposure to coal mine dust. Fine dust particles, smaller than five microns, enter the small airways and air sacs in the lungs, blocking air passages and causing lesions that contain coal dust, immune cells called macrophages, and fibroblasts (cells that produce connective tissue). If silica is present in the dust, it can release substances that worsen lung damage.

The CWP has two forms: the simple form and progressive massive fibrosis (PMF). The simple form causes small lung scars, usually less than 10 mm in size, and can lead to symptoms such as a cough or shortness of breath. However, some people with simple CWP may not show any symptoms. In contrast, PMF results in larger scars, greater than 10 mm, and can cause more severe symptoms, such as difficulty in breathing, black sputum (coughing up dark-colored mucus), chronic cough, pulmonary hypertension (high blood pressure in the lungs), frequent lung infections, and heart problems (Han et al. 2018).

4.3.3 Asbestosis and Other Related Diseases

Asbestosis is a serious lung disease caused by inhaling asbestos fibres, which are tiny, harmful strands found in certain rocks and soil. These fibres are classified into two main types: serpentine and amphibole. Serpentine fibres, such as chrysotile, are curly and flexible, so they tend to stay in the upper parts of the respiratory system where the body can remove them more easily. In contrast, amphibole fibres, like crocidolite and amosite are straight and stiff, making them more toxic. These fibres can penetrate deeper into the lungs, causing more severe health problems. When inhaled, asbestos fibres lead to scarring in the lungs, making it difficult to breathe. Over time, this scarring worsens, causing symptoms like shortness of breath, coughing, and chest tightness. Doctors must differentiate asbestosis from other conditions that also cause lung scarring, such as idiopathic pulmonary fibrosis (a disease where lung tissue is damaged for unknown reasons) and rheumatoid arthritis (an autoimmune disease that can affect the lungs). This process of distinguishing between diseases is known as differential diagnosis (Bhandari 2022). Exposure to asbestos is the main cause of most mesotheliomas (a rare cancer affecting the thin membranes around the chest and abdomen). Asbestos exposure can also lead to cancers of the lungs, larynx (voice box), and ovary (NCI 2022). Due to the severe health risks of asbestos, strict regulations are in place to limit exposure, especially in industries like construction, where asbestos is commonly found (Fig. 4.1).

4.3.4 Mercury Poisoning

Artisanal and Small-Scale Gold Mining (ASGM) is a major source of mercury emissions worldwide. This mining method is typically carried out by individuals or small groups with limited tools. In ASGM, miners use mercury (a toxic heavy metal) to extract gold from ore by mixing it with the material to form an amalgam (a mixture of mercury and gold). The amalgam is then heated, causing the mercury to evaporate, leaving behind the gold. However, this process releases mercury vapours (gaseous form of mercury) into the air, which poses serious health risks to miners and nearby communities due to inhalation. Mercury exposure can be harmful, and its effects

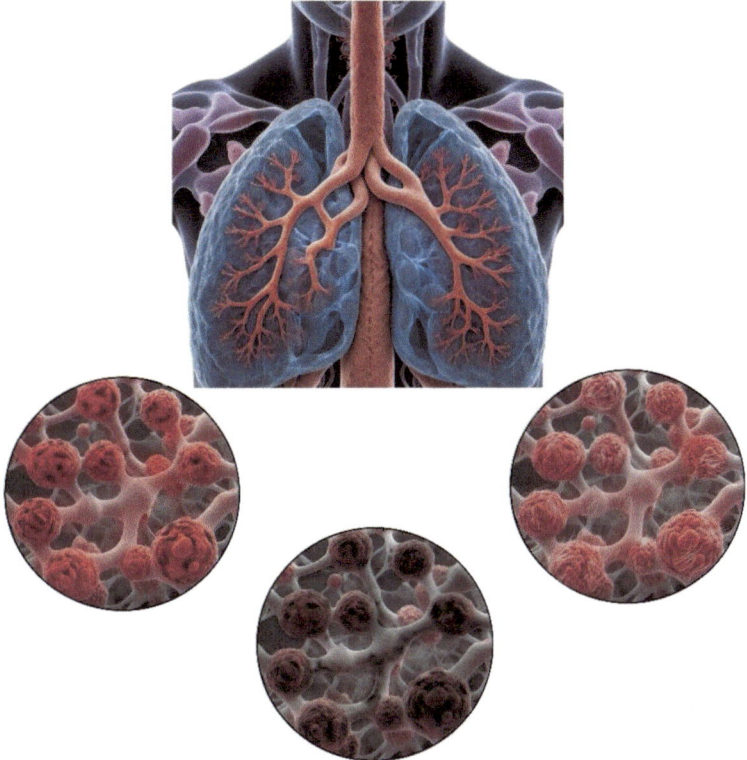

Fig. 4.1 Sketch showing Silicosis, Coal workers' pneumoconiosis (CWP), and Asbestosis from left to right

depend on the type of mercury, the amount absorbed, and the way it enters the body. When miners heat the amalgam, they inhale mercury vapour, which is absorbed mainly through the trachea (airway passages leading to the lungs). Once inhaled, around 67–87% of mercury vapour enters the body. Skin contact and swallowing play a much smaller role in absorption. Mercury can accumulate in body tissues, and different forms of mercury affect the body in various ways. Inhaled mercury crosses into the bloodstream (the network of blood vessels) and can reach critical areas like the brain and placenta (the organ that nourishes a baby during pregnancy), leading to symptoms such as tremors (involuntary shaking), memory loss, and mood changes. In severe cases, it can cause lung and kidney damage. Chronic exposure to mercury vapour, especially in ASGM, can result in long-term health problems like neurological disorders (brain and nervous system issues), gum disease (infection or inflammation of the tissues surrounding the teeth), gum discolouration, and kidney problems. Detection of mercury in the body can be done through tests on blood, urine, or even hair samples. Acute poisoning (short-term exposure) can cause symptoms

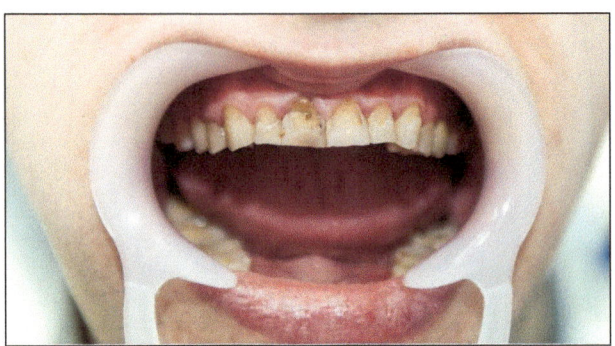

Fig. 4.2 Image showing dental fluorosis

like coughing, breathing difficulties, and even respiratory failure (when the lungs can't supply enough oxygen) (Cicek-Senturk et al. 2014; Rowens et al. 1991).

4.3.5 Fluorosis and Associated Diseases

Fluorosis is a serious public health issue caused due to excessive exposure to fluoride, often found in contaminated drinking water, especially in mining areas. The most common type is dental fluorosis, which causes changes in the appearance of teeth, including discoloration (colour changes), white spots, and, in severe cases, brown stains and pitting (small holes). These changes can lead to social stigma and reduced self-esteem. Chronic exposure to high levels of fluoride can also cause skeletal fluorosis, which results in joint pain, stiffness, and changes in bone structure that can limit movement. In addition to affecting teeth and bones, fluorosis can impact brain function, especially in children, and may interfere with hormone levels, potentially leading to thyroid issues (problems with the gland that regulates metabolism) and reproductive problems (DenBesten and Li 2011) (Fig. 4.2).

4.3.6 Berylliosis and Associated Diseases

Beryllium exposure at work, especially in industries like mining and manufacturing, can lead to serious health issues. In the United States, safety regulations indicating tolerance limit of beryllium exposure to mine workers, commonly known as the Permissible Exposure Limit or PEL, have reduced cases of severe lung disease called acute beryllium pneumonitis (a type of lung inflammation caused by inhaling beryllium dust or fumes). Despite these regulations, deaths from cancers of the windpipe (tracheal cancer), airways (bronchial cancer), and lungs (lung cancer) linked to workplace exposure have increased globally. This condition is similar to chemical pneumonitis (lung inflammation caused by chemicals) and can result in bronchiolitis (inflammation of the small airways in the lungs), pulmonary edema (fluid buildup in the lungs), and pneumonitis (general lung inflammation). Beryllium exposure can also affect the skin, causing irritation, sores (ulceration), or small lumps under the skin called subcutaneous granulomas (small lumps of inflamed tissue). Chronic beryllium disease (CBD), also called berylliosis, is a long-term lung disease caused by beryllium exposure. It occurs when the immune system reacts to beryllium, forming granulomas (small lumps of inflamed tissue) in the lungs. This disease is like sarcoidosis (a disease that also causes granulomas). While CBD can remain stable for many years, it sometimes progresses to pulmonary fibrosis (scarring of the lungs) and increases the risk of lung cancer (Sizar et al. 2023; Mayer and Hamzeh 2015).

4.3.7 Radon-Induced Lung Cancer and Associated Diseases

Radon-222 is a radioactive gas that forms when uranium-238 in the soil decays. It is often found in areas with rocks such as granite, gneiss, and schist, and the amount of radon released depends on soil type and weather conditions (Cinelli et al. 2019). Radon can enter buildings through cracks and gaps in foundations, particularly in areas with high uranium levels (WHO 2009). The main health risk from radon exposure is the lung cancer. As radon-222 decays, it releases alpha radiation, a type of ionizing radiation that can damage DNA (the molecule carrying genetic information) in lung cells. This damage can cause mutations (changes in DNA) that lead to cancer. Common symptoms of radon-related lung cancer include a persistent cough, shortness of breath, and chest pain, while advanced stages may involve hemoptysis (coughing up blood), weight loss, and fatigue (Riudavets et al. 2022). Since radon exposure typically doesn't cause symptoms in its early stages, testing indoor radon levels is essential to prevent long-term health risks. When radon is inhaled, radioactive particles settle in the bronchial tubes (airways) and alveoli (small air sacs in the lungs). While some particles are expelled through sneezing or coughing, others are removed by mucociliary clearance, a process where mucus traps particles and removes them from the lungs. However, if radon particles remain in the lungs, they can damage cells over time and increase the risk of lung cancer (Degu et al. 2021).

4.3.8 Manganism and Associated Diseases

Manganism is a disease caused by breathing in manganese dust or fumes over an extended period, particularly in occupations like mining or welding. Manganese is a metal found in certain rocks and minerals. When inhaled, it can accumulate in the brain, particularly in the basal ganglia (a part of the brain that controls movement). This buildup can lead to nerve damage and cause symptoms like fatigue, weakness, headaches, and tremors (shaking). As the condition worsens, it can result in more severe issues, such as difficulty in walking, muscle stiffness, and tremors, which are similar to Parkinson's disease (a neurodegenerative disorder that affects movement). Long-term exposure to manganese can cause permanent brain damage. Workers in environments where manganese dust is prevalent are at a higher risk. Using protective gear and following safety regulations is crucial in preventing manganism (Lucchini et al. 2009).

4.3.9   Blue Baby Syndrome (Methemoglobinemia) Due to Nitrate Contamination

Using nitrate-contaminated water to prepare infant formula can lead to Blue Baby Syndrome (methemoglobinemia), a serious condition in babies. Infant formula is a liquid food that provides nutrients when breastfeeding is not possible. Babies with this condition may have a blue-grey skin colour and may feel irritable or tired. If not treated quickly, it can get worse and may lead to coma or death (Knobeloch et al. 2000). Nitrate contamination in water is often caused by excessive fertilizer use and the buildup of organic matter in soil, which increases nitrate levels (Wick et al. 2012). Nitrates are not harmful by themselves, but in the body, they turn into nitrite (NO_2^-). Nitrite then affects haemoglobin (the part of blood that carries oxygen), turning it into methemoglobin (MHb), which can't carry oxygen properly. This causes a lack of oxygen in tissues, making the skin turn blue (cyanosis). In some cases, methemoglobinemia can be worse if a baby has an enzyme problem or other conditions that affect haemoglobin (Al-Absi 2013).

4.4   Mining and Associated Occupational Health Hazards

Mining provides essential resources but also creates significant health risks for miners and nearby communities. Miners are exposed to dust, harmful chemicals, and physical injuries. In some regions, accidents in mining are frequent, especially during the transportation of materials or manual labour (Onder and Mutlu 2016). Manual labour involves physically demanding tasks that use hands and muscles. Common injuries include fractures (broken bones) and back pain. Ergonomic issues, such as muscle strain from poor posture or repetitive tasks, are also widespread in small-scale mining (Jiménez-Forero et al. 2015; Kyeremateng-Amoah and Clarke 2015). Small-scale mining operations often lack advanced technology, which makes safety concerns difficult to manage. Miners, especially those working with rock and sand, are at high risk of developing lung diseases if proper safety measures are not followed (Steen et al. 1997; Kistnasamy et al. 2018). Migrant miners, who often do not have access to health checkups or adequate protection, are especially vulnerable. Mining accidents, like falling or getting caught in equipment, are also common and can cause

Fig. 4.3 Sketch showing an overview of the mining activities

serious injuries or even death. Fatigue, caused by long hours of work, makes accidents more likely by dwindling alertness (Wesdock and Arnold 2014). The focus on profit in the mining industry sometimes leads to compromising safety and working conditions. Therefore, safety measures are necessary to prevent accidents and protect workers' health. Researchers have long called for better safety measures in mining (Donoghue 2004). Today, ensuring safe mining practices balancing worker's health and well-being with the efficiency and profitability of the industry is crucial (Maier et al. 2014; Woodward and Hales 2014) (Figs. 4.3 and 4.4).

4.4.1 Physical Hazards in Mining

4.4.1.1 Respiratory and Lung Hazards

Mining operations expose workers to dust particles that can cause serious respiratory diseases. These dust particles come in two types: inhalable (INH) and respirable (RES). INH dust consists of larger particles that settle in the nose and throat, while RES dust has smaller particles that reach the lungs. These particles, usually between 1 and 100 μm in size, are created during activities like blasting and drilling. Breathing this dust over time can lead to various health problems, including asthma (narrowing of airways), tracheitis (inflammation of the windpipe), pneumonia (lung infection), allergic rhinitis (hay fever), chronic bronchitis (long-term airway inflammation), silicosis (lung disease from silica dust), tuberculosis (lung infection), emphysema

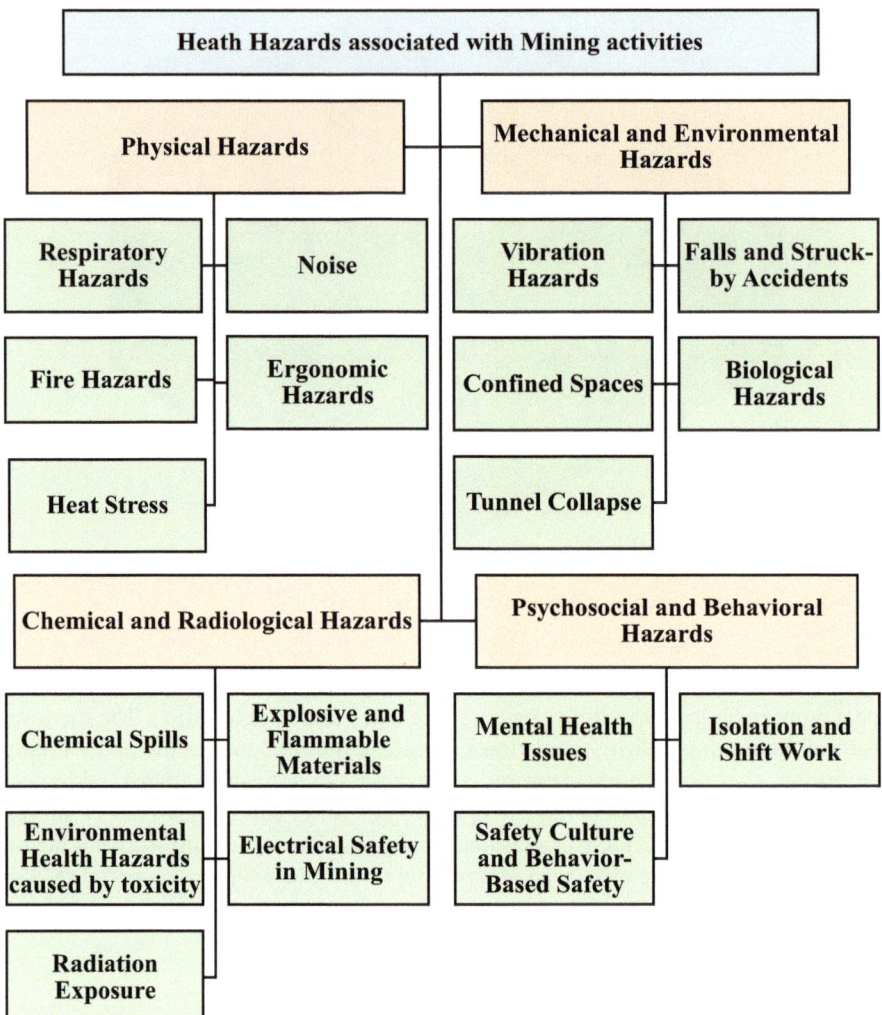

Fig. 4.4 Showing types of hazards associated with mining operations

(damage to lung air sacs), kidney failure, and cancer. To protect workers, safety agencies set limits on dust exposure. The Occupational Safety and Health Administration (OSHA) limits respirable dust to 5 mg/m^3 over an 8-h workday. The American Conference of Governmental Industrial Hygienists (ACGIH) recommends keeping respirable dust levels below 3 mg/m^3 and inhalable dust below 10 mg/m^3 (Rumchev et al. 2022). These guidelines help reduce the risk of health problems from long-term dust exposure in mining environments.

Inhalable Dust:

Inhalable dust particles, typically larger than 10 μm in diameter, pose a significant risk to miners' respiratory health. These particles are inhaled through the nose and mouth, causing irritation, coughing, and discomfort. While they do not reach as deep into the lungs as smaller respirable dust particles, long-term exposure to inhalable dust can still lead to respiratory issues. This type of dust is commonly produced during mining activities like blasting, drilling, and handling materials (Aeroqual 2022).

Respirable Dust:

Respirable dust, consisting of fine particles smaller than 10 μm, is a major concern in mining. These tiny particles can penetrate deep into the lungs, leading to severe respiratory diseases. Prolonged exposure to respirable dust can result in chronic bronchitis, where the airways become inflamed, causing persistent coughing and mucus buildup. It can also increase the risk of lung cancer, a serious and potentially fatal disease where abnormal cells grow uncontrollably in the lungs. Mining activities like drilling, blasting, and crushing rocks produce these fine particles. To protect miners, it is crucial to follow strict dust control measures and wear respiratory protection to reduce risk of these health issues (Shekarian et al. 2021).

4.4.1.2 Noise Hazards

Noise pollution in mining areas is a serious problem, especially in India, where it has been an ongoing issue for many years (Kulkarni and Mandal 2015). Miners are exposed to loud sounds from machines and operations, which can harm their health and affect nearby communities. Some health problems caused by noise include tinnitus, ringing or buzzing in the ears, and noise-induced hearing loss (NIHL), which occurs when loud sounds damage the inner ear and lead to permanent hearing loss. Noise can also make it hard to focus, sleep, or communicate, and can cause stress. Because of these risks, the Indian government officially recognized NIHL as a disease under the Mines Act in 2011 (Manwar et al. 2016; Thakkar et al. 2022).

4.4.1.3 Ergonomic Hazards

Ergonomic injuries, also known as repetitive strain injuries (RSIs) or musculoskeletal disorders (MSDs), happen when the body is exposed to repetitive physical stress without enough rest. These injuries can affect muscles, bones, tendons (the tissues connecting muscles to bones), and nerves. Common examples include carpal tunnel syndrome (pain in the hand and wrist), bursitis (swelling in the joints), and tendinitis (inflammation of tendons from overuse) (Salve et al. 2022). Back injuries are the most common and costly, especially in jobs that require heavy lifting or repetitive

movements. Symptoms can include pain, stiffness, numbness, weakness, and difficulty moving, particularly in the back, arms, wrists, neck, and shoulders (Yassi 1997).
Ergonomic hazards are factors in the workplace that put strain on muscles and joints,
such as repetitive movements, awkward positions, or overexertion. These hazards
can lead to injuries if not properly managed. To prevent such issues, it's important
to design workplaces better, use tools correctly, and take regular breaks (Salve et al.
2022).

4.4.1.4   Heat Stress

At greater depths, heat inside underground mines increase due to factors such as
the natural rise in temperature with depth (geothermal gradient), the use of large
machinery, heat from exposed rock surfaces, compression of air, and the body's
metabolism. Prolonged work in such hot environments can negatively impact miners'
health, safety, and efficiency (Roy et al. 2022). Heat-related illnesses are common
in these conditions and can range from minor to life-threatening if not managed
properly. Prickly heat, also known as heat rash, causes an itchy skin condition due to
blocked sweat ducts. Fainting, or heat syncope, occurs after long periods of standing
or working in the heat. Muscle cramps can result from a loss of salts (electrolytes)
through sweating, particularly in those not used to high temperatures. Heat exhaustion
is more severe and includes symptoms such as fatigue, nausea, clammy skin, and an
elevated body temperature. If untreated, it can progress to heat stroke, a medical emergency with symptoms like confusion, convulsions (seizures), or unconsciousness,
along with hot and dry skin (Resource Safety and Health Queensland 2023).

4.4.1.5   Fire Hazards

For a fire to start, three key elements must be present: fuel (anything that can burn),
oxygen (or an oxygen-rich compound), and an ignition source (like heat or some-
thing above the fuel's flash point). When these elements are in the right amounts,
a fire can start and continue by a chemical chain reaction. Fires are common in
mining, and outbreaks are frequent. Risks include fuel sources such as debris, wood,
flammable liquids, gases, combustible metals, and electrical equipment. Ignition
sources can be hot surfaces, cigarette butts, electrical sparks, lightning, mechanical
actions like cutting or grinding, or chemical reactions. Mining areas, whether in
exploration, underground, or remote locations, are particularly dangerous because
of limited access to firefighting resources, tight spaces, restricted evacuation routes,
and external threats like bushfires. To reduce risks, it's important to identify fire
hazards, keep fuel and ignition sources apart, maintain equipment, and have strong
risk and emergency plans (Mines Safety and Inspection Act 1994) (Fig. 4.5).

Fig. 4.5 Sketch showing different types of physical hazards occurring in mine. **a** Respiratory hazards, **b** Noise hazard, **c** Ergonomic hazard, **d** Heat stress, **e** Fire hazrds

4.4.2 Mechanical and Environmental Hazards

4.4.2.1 Vibration Hazards

The body reacts to vibration in different ways depending on which part is affected. There are two main types of vibration: whole-body vibration (WBV) and segmental vibration. WBV affects the whole body through the surface it touches, while segmental vibration, like hand-arm vibration (HAV), targets specific areas, such as the hands. Each type needs different methods for study and control, as they cause different health problems. Vibration, a back-and-forth movement, is known to cause health issues, especially for miners using tools like jackhammers. Workers exposed to HAV from vibrating tools may develop conditions such as vibration-induced white finger (VWF) or hand-arm vibration syndrome (HAVS). These conditions were first identified in 1911 among stone cutters using air-powered hammers. Symptoms include tingling, numbness, and pale fingers, especially in cold weather or among smokers. In severe cases, the damage can be permanent, and sometimes amputation may be necessary (Mandal et al. 2006).

4.4.2.2 Confined Space Hazards

Confined spaces in mining pose serious dangers to workers' safety. These hazards include toxic atmospheres, low or high oxygen levels, flammable or explosive air, moving liquids or loose solids, and extreme heat. Toxic atmospheres from harmful substances can impair breathing, cause unconsciousness, or even be fatal. Low oxygen levels occur when other gases replace oxygen, while too much oxygen increases the risk of fire or explosion. Flammable air can ignite and lead to explosions. Moving liquids or loose solids, such as sand or grains, can cause drowning or suffocation. Extreme heat, especially with poor ventilation or inadequate protective gear, can result in heat-related illnesses.

4.4.2.3 Struck by, Caught Between and Fall Hazards

"Struck by" and "caught between" accidents are serious risks in mining workplaces. A "struck by" accident occurs when a worker is hit by a moving object, like a vehicle or equipment. A "caught between" accident happens when someone is trapped between two objects, like a machine and a wall. Moving vehicles, such as forklifts and construction equipment, can be dangerous if not used properly. Workers driving these vehicles need to check the vehicle before use, drive carefully, and avoid overloading. Workers near these vehicles must stay alert, especially near blind spots, where the driver cannot see them. Falling objects are another risk. If materials are not stacked properly, they can fall and injure people below. Falls are also a big problem in mining, making up about 30% of all accidents. Falls can happen on the same level, like tripping or slipping, or from a height, like falling off a ladder. These accidents often cause sprains (stretched or torn ligaments), strains (damage to muscles or tendons), or fractures (broken bones), especially in the back, shoulders, and knees.

4.4.2.4   Tunnel Collapse

Tunnel collapses, also known as roof falls, are a significant hazard in mining, often causing severe injuries or even fatalities (Duzgun and Einstein 2004). These occur when the roof or walls of an underground tunnel suddenly give way, releasing rock and debris (Barton 2016). Factors contributing to collapses include the stability of the surrounding rock, which may contain natural cracks or faults, and the mining method used. If roof support systems are inadequate or excavation is done improperly, the risk of collapse increases. Changes in rock stress during mining can also contribute to these events. When a tunnel collapses, miners can be trapped or buried under rock and debris, leading to injuries such as broken bones, head trauma, or internal injuries. The dust from the collapse can cause respiratory issues, making it difficult to breathe, especially if miners are trapped for extended periods. This can lead to long-term lung problems. Besides physical injuries, tunnel collapses can cause psychological distress, including panic and trauma, as miners face life-threatening situations. In addition, collapses can block escape routes, delaying rescue efforts, and prolonged entrapment can result in dehydration, exhaustion, and other complications. These events also pose risks to rescuers, who must work in dangerous and unstable conditions to save trapped miners (Mahdi et al. 2022).

4.4.2.5   Biological Hazards

In remote mining areas, tropical diseases like malaria and dengue fever are common due to the prevalence of mosquitoes. Malaria causes symptoms such as fever, chills, and fatigue, while dengue fever leads to severe headaches, joint pain, and rashes. Mines can also face other health risks like leptospirosis, a bacterial infection caused by exposure to water or soil contaminated with rat urine, and ancylostomiasis, a hookworm infection that affects the intestines and can lead to anaemia. While these diseases have been largely controlled in developed countries through measures like pest control and improved sanitation, they remain a risk in more remote regions. In addition, many mining sites use cooling towers, which need regular testing for bacteria like Legionella. This bacterium can cause Legionnaires' disease, a severe lung infection. To protect workers, it's vital to keep water sources clean and free from harmful microorganisms, ensuring a safe work environment (Donoghue 2004) (Fig. 4.6).

Fig. 4.6 Sketch showing mechanical and environmental hazards (**f** Vibration hazards, **g** Confined space hazards, **h** Struck by hazards, **i** Biological hazards, **j** Tunnel collapse)

4.4.3 Chemical, Radiological and Electrical Hazards

4.4.3.1 Chemical Spills

Chemical spills in mining can severely harm both the environment and human health. These spills typically occur due to accidents, equipment failures, or improper handling and storage of hazardous substances. Common chemicals involved include acids, heavy metals, cyanide, and other toxic materials used in mining processes such as extraction and waste management. When a spill happens, it can contaminate the soil, rivers, lakes, and groundwater, impacting plants, animals, and ecosystems. The effects of chemical spills can persist for a long time, as these substances may remain in the environment and accumulate in the food chain. Even after a spill is cleaned up, harmful chemicals can still be present in plants, animals, and humans. Cleaning up such spills is challenging and costly, requiring specialized knowledge, equipment, and significant resources. In addition to environmental damage, nearby communities may face health problems, job losses, and social disruption (Graczyk et al. 2021).

4.4.3.2 Explosive and Flammable Materials

Explosions and flammable hazards are significant safety concerns in mining. Explosives are commonly used to break rocks and extract minerals (Bajpayee et al. 2004). However, if not handled, stored, or used correctly, they can cause dangerous explosions and fires. In addition to explosives, materials like fuels, gases, and dust in mining areas can also ignite or explode, posing risks to both miners and nearby communities (Lin et al. 2019). Gas explosions, especially in coal mining, are particularly hazardous. They occur when gases, like methane, mix with air and ignite, typically due to an electric spark or heat source. The tunnels and pathways in mines can amplify these explosions, turning calm gas flows into fast-moving waves, called turbulent flow. This leads to pressure waves, which are forceful air movements that can cause injuries, including ear damage, lung issues, or structural collapses. To protect miners, strict safety measures and advanced methods are essential for preventing and controlling gas explosions (Dou et al. 2023).

4.4.3.3 Radiation Exposure

Uranium mining presents unique health risks due to exposure to radionuclides (radioactive atoms). These risks come from both internal exposure (through inhalation, ingestion, or skin absorption) and external exposure to beta and gamma radiation. Alpha particles, emitted by some radioactive substances, are harmful if inhaled as they cause significant damage to nearby cells through ionization. Beta particles penetrate deeper into tissues than alpha particles but cause less damage. Gamma rays, being highly penetrating, pose an external threat due to their strong radiation. Prolonged exposure to radiation increases the risk of developing cancer. In uranium mining, the radioactive materials of concern come from the uranium and thorium decay series, a process that releases radiation (Committee on Uranium Mining in Virginia 2011).

4.4.3.4   Electrical Safety in Mining

Electricity is vital in mining for powering equipment like pumps, trucks, lights, and ventilation fans. However, it also poses a risk of electric shock, which triggers when someone accidentally becomes part of an electrical circuit, such as touching a bare wire, faulty equipment, or exposed electrical connections. To prevent shocks, miners should wear protective gear like rubber boots and gloves, and work on insulated surfaces. The human body conducts electricity well due to its water content. The severity of an electric shock depends on the strength of the current. For example, a current of 1 mA is barely felt, while 100–200 mA can cause serious heart problems or death. Electric burns and arcs can also cause injuries. To stay safe, miners should check their protective gear regularly, secure electrical connections, use insulated tools, and always lock out equipment before working on it. Avoiding live wires and wearing insulated gear are key steps to reduce electrical risks (Fig. 4.7).

Fig. 4.7 Sketch showing **k** Chemical hazards, **l** Flammable hazards and **m** Electrical hazards

4.4.4 Psychosocial and Behavioural Hazards

4.4.4.1 Mental Health Issues

Mining work can lead to mental health issues. These problems are influenced by four main factors. First, personal factors like a miner's personality and personal life can affect their mental health. For example, having a positive attitude can help, while substantial abuse can make things worse. Second, the work environment is stressful. Dangerous conditions, long hours, and the pressure of balancing work and family life can lead to anxiety or depression. Third, physical health problems, like back pain, are common in mining and can cause extra stress. Job dissatisfaction can also lower mental health. Finally, organizational factors, like unsafe conditions and difficult relationships with coworkers or managers, can increase stress. All these factors affect miners' mental health; therefore, it is important to create better ways to support them (Matamala et al. 2021).

4.4.4.2 Isolation and Shift Work

Shift work and isolation in mining can cause many health problems. Many miners feel sad or anxious because they don't get enough sleep or have trouble spending time with family and friends. Around 75% of shift workers have trouble sleeping, which makes them very tired. This tiredness can lead to stress and make their lives harder. On the physical side, shift workers are more likely to get heart problems, like high blood pressure, because their sleep and eating habits are not regular. They can also get stomach problems, like ulcers, because of eating at strange times and feeling stressed. Miners who work shifts also often have pain in their backs, necks, and shoulders from the physical work and lack of rest. To help reduce these problems, it's important to set good work schedules, limit how many hours people work, educate them about staying healthy, and offer support for mental health (Omidi et al. 2017).

4.4.4.3 Safety Culture and Behaviour-Based Safety

Safety attitude, peer influence, safety knowledge, and risk perception are key to keeping miners safe. Safety attitude is how workers feel about following safety rules and avoiding accidents. In places where workers have low education, safety culture is often weak, leading to bad attitudes, poor knowledge, bad influences from coworkers, and not understanding risks. It's important for miners to have basic safety knowledge, like how mining works, how to use equipment safely, and how to follow safety rules. When miners are trained and informed, they are more confident and better able to handle dangers (Wu et al. 2017) (Fig. 4.8).

Fig. 4.8 Sketch showing emotional distress of mine worker

4.5 Geohazards and Associated Diseases

Geohazards are natural events that can harm people, property, and the environment
(Komac and Zorn 2013). These include earthquakes, volcanic eruptions, landslides,
floods, tsunamis, sinkholes, and avalanches. They happen suddenly, making it hard to
predict them and therefore cause severe damage of property and loss of life (Bokwa
2013). Geohazards are caused by Earth's internal processes, such as tectonic movements, or surface changes like erosion. While these events cause immediate damage,
they also induce long-term health issues. After such an event, disease outbreaks are
common due to disruptions in the environment. For example, earthquakes and floods
can damage water systems, leading to contamination. This can cause an outbreak of
diseases like cholera, typhoid fever, and hepatitis A (Ramírez-Castillo et al. 2015).
When infrastructure collapses, it results in poor sanitation and overcrowded shelters,
increasing the spread of gastrointestinal and respiratory diseases (Mavrouli et al.
2023). Floods and tsunamis leave behind stagnant water, which is a breeding ground
for mosquitoes that spread malaria and dengue fever (Coalson et al. 2021). Additionally, volcanic eruptions release ash and gases, which can damage the lungs and

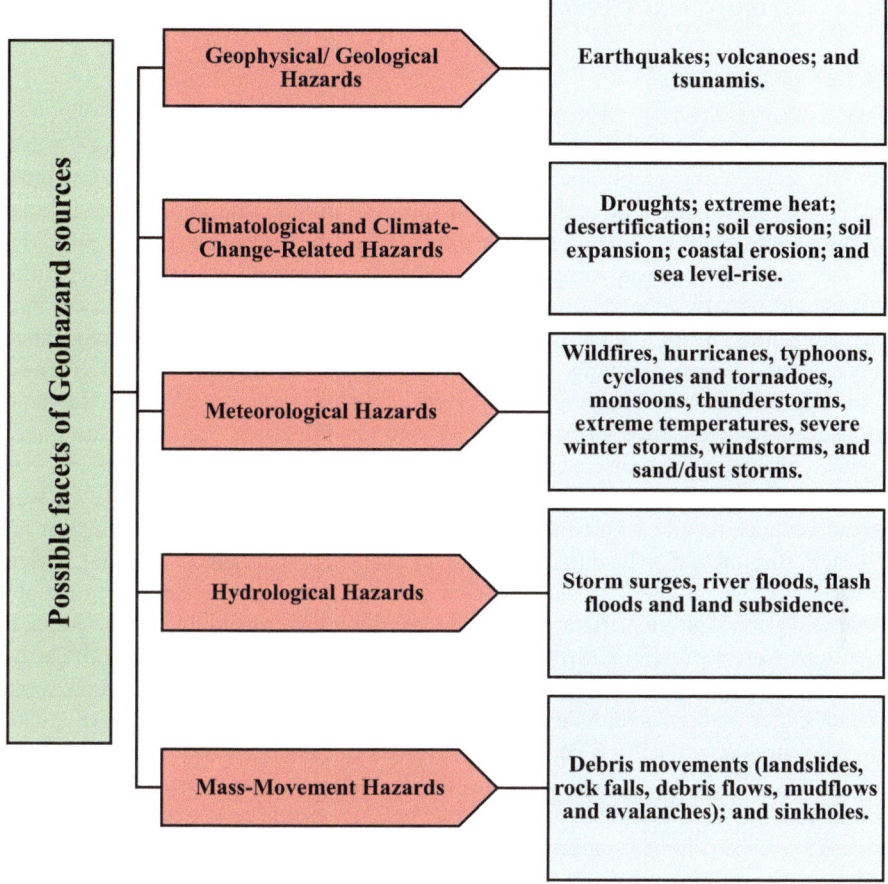

Fig. 4.9 Possible sources of geohazards (modified after Randive 2013)

worsen health problems (Mueller et al. 2020). Understanding the hazard potential of the geological events helps us recognizing the type of health hazard it can cause. Although geohazards are rare, they happen suddenly and create widespread damage that requires quick response (Alaska Nature and Science 2019) (Fig. 4.9).

4.5.1 Geophysical/Geological Hazard

4.5.1.1 Earthquakes

An earthquake is a sudden shaking of the ground caused by the movement of Earth's tectonic plates along a fault line. The damage depends on the earthquake's strength, the type of ground, the time of day, and the building strength. Earthquakes can cause building damage, fires, landslides, liquefaction (when the ground behaves like liquid), and tsunamis (large ocean waves) (Earle 2019). The severity of injuries and deaths from an earthquake depends on the local geology, the earthquake's magnitude, and conditions like time of day and weather (Mavrouli et al. 2023). Survivors may experience mental health problems such as post-traumatic stress disorder (PTSD) and major depression (MD) (Salcioglu et al. 1999). In addition to mental health impacts, diseases can spread after an earthquake. Common infections include pneumonia, stomach problems, skin diseases, and wound infections (Mavrouli et al. 2023). For instance, after the 1994 Northridge earthquake in California, the valley fever was spread because people inhaled spores of the Coccidioides fungus (Schneider et al. 1997). Earthquakes also lead to an increase in diseases like tuberculosis (TB), which spreads in crowded shelters, and stomach infections like diarrhoea, due to unsafe water and poor sanitation. After the 2016 Kumamoto earthquake in Japan, stomach infections were reported (Yorifugi et al. 2016). Similarly, after the 1999 İzmit earthquake in Turkey, diseases caused by bacteria like Shigella and Salmonella were spread. Other outbreaks include cholera in Nepal after the 2015 Gorkha earthquake and Zika virus after the 2016 Ecuador earthquake. In Pakistan's 2005 Kashmir earthquake, wound infections from bacteria like *Pseudomonas* and *Enterobacter* were common. After the 2008 Sichuan earthquake in China, some people developed infections from Clostridium bacteria, which can cause gas gangrene, a life-threatening condition (Mavrouli et al. 2023) (Fig. 4.10).

4.5.1.2 Volcanic Eruption

Volcanic eruptions can cause many health problems, sometimes more than other natural disasters. These effects can happen nearby or even thousands of kilometres away due to the spread of volcanic gases and ash. Volcanic eruptions can also affect the global climate (Hansell et al. 2006). Tephra, which includes ash particles smaller than 2 mm, can cause breathing problems when inhaled, skin and eye irritation, and even building collapse if too much ash builds up on roofs. It can also harm crops and vegetation. Tephra may contain toxic fluorine from volcanic gases, which can poison animals if they graze on ash-covered land. It can also contaminate water, though human poisoning is rare (Witham et al. 2005). Ash clouds can damage airplanes by affecting jet engines and cockpit windows, though no crashes have occurred so far (Hansell et al. 2006). One of the most dangerous volcanic hazards is pyroclastic density currents (PDCs)—fast-moving flows of hot ash, gas, and rocks that can reach

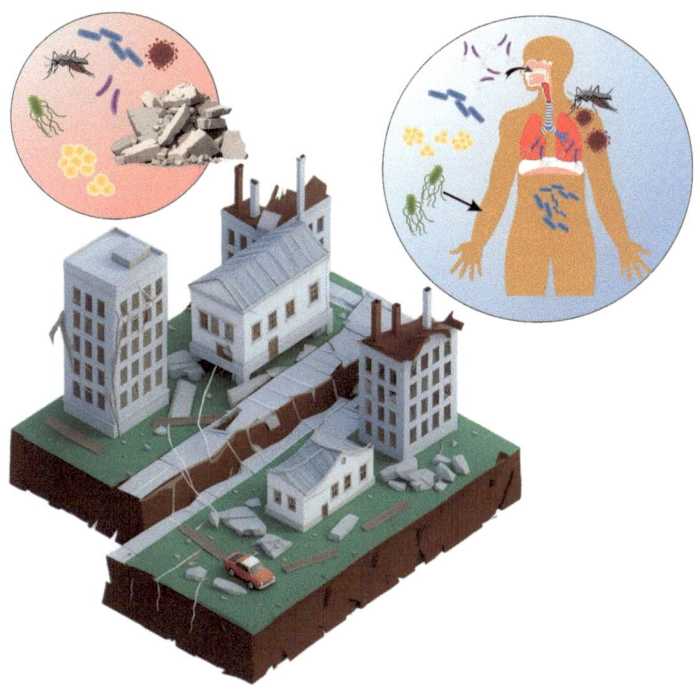

Fig. 4.10 A thematic sketch showing health hazard/destruction due to earthquake. The bigger circle shows possible pathways to enter the body whereas, the smaller circle shows potential pathogens and other contaminants

speeds of 350 km/h and temperatures of several hundred degrees Celsius. PDCs can cause death by heat shock, asphyxiation (difficulty breathing), lung damage, and severe burns. The best way to stay safe is to evacuate the area (Baxter 1990; Mastrolorenzo 2001). Volcanoes can also cause landslides, triggered by earthquakes or heavy rain. Even small landslides can cause injuries and property damage, as seen in the 1998 landslide at Casita volcano. On volcanic islands, landslides can cause tsunamis if debris flows into the sea (Hansell et al. 2006). Another deadly hazard is lahars, fast-moving mixtures of water and volcanic debris. These can form from snowmelt, heavy rain, or the sudden release of water from volcanic lakes. Lahars can travel over 50 km/h and cover long distances, like the 1985 Nevado del Ruíz eruption in Colombia, which caused a lahar that killed around 22,800 people (Baxter 1990). Volcanic gases, such as carbon dioxide (CO_2), sulfur dioxide (SO_2), and hydrogen sulfide (H_2S), can cause serious health problems. CO_2 can lead to asphyxiation (lack of oxygen), while SO_2 and H_2S can cause respiratory issues and irritation of the eyes and throat. Large eruptions have been linked to increased respiratory illnesses and deaths (Hansell et al. 2006). A famous example of volcanic devastation is the eruption of Mount Vesuvius in 79 AD, which buried the city of Pompeii under ash and pumice. Thousands of people died, many from asphyxiation due to inhaling

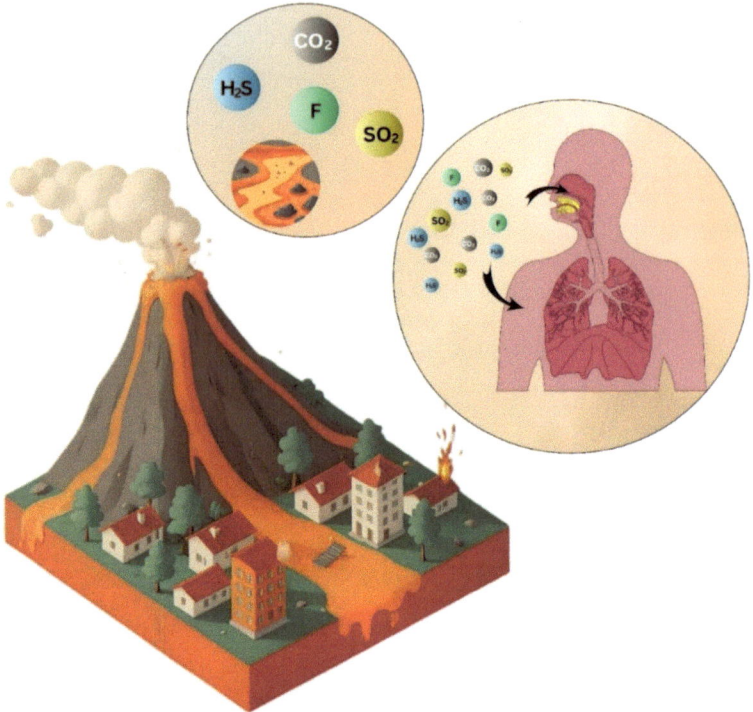

Fig. 4.11 A thematic sketch showing health hazard/destruction due to volcanic eruption. The bigger circle shows possible pathways to enter the body whereas, the smaller circle shows potential pathogens and other contaminants

volcanic ash (Hansell and Oppenheimer 2004). Large eruptions, like the Laki fissure eruption, have caused widespread health impacts, increasing respiratory diseases and deaths (Baxter et al. 1986; Witham and Oppenheimer 2004) (Fig. 4.11).

4.5.1.3   Tsunamis

A tsunami, meaning "large harbour wave" in Japanese, is a series of large waves caused by a sudden vertical movement of water. Tsunamis are usually triggered by underwater earthquakes but can also result from volcanic eruptions, landslides, or meteorite impacts (Doocy et al. 2013). Although tsunamis are rare, they can cause widespread destruction and affect many people when they occur (Noji 1997). Injuries from tsunamis are often due to blunt force trauma or puncture wounds, caused by debris like concrete blocks, trees, or sharp objects. After the 2004 tsunami in Aceh, many survivors suffered from respiratory infections, especially from inhaling seawater, which led to a condition known as "tsunami lung". This condition is caused

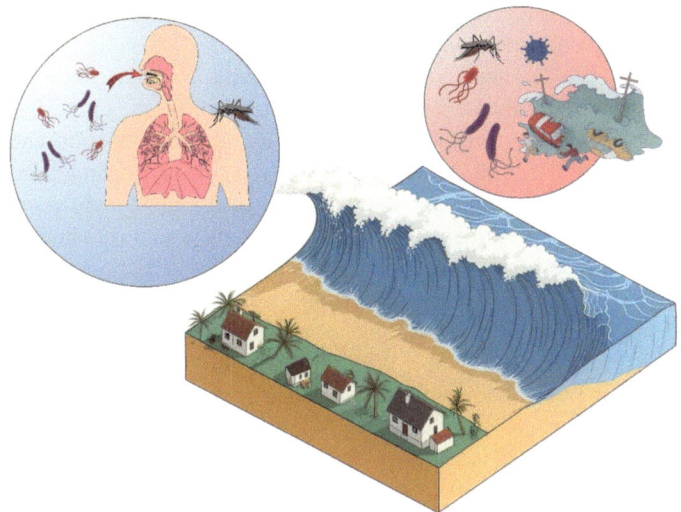

Fig. 4.12 A thematic sketch showing health hazard/destruction due to tsunami. The bigger circle shows possible pathways to enter the body whereas, the smaller circle shows potential pathogens and other contaminants

by bacteria in seawater, such as *Aeromonas* and *Pseudomonas*. In some cases, infections were caused by rare and antibiotic-resistant bacteria. Tsunami-related injuries are often contaminated with soil and debris, which can introduce harmful bacteria like *Pseudomonas aeruginosa*, *Stenotrophomonas maltophilia*, and *Klebsiella pneumoniae.* These bacteria can lead to unusual infections. Additionally, soil-contaminated wounds caused an increase in tetanus cases, though the number of cases returned to normal within a month. Proper identification of these pathogens and their resistance to antibiotics is crucial for effective treatment, which requires proper diagnostic tools. Despite concerns, the 2004 Indian Ocean tsunami did not cause outbreaks of diseases like malaria, measles, cholera, or dengue. In fact, malaria cases in Aceh province significantly dropped after the tsunami (Keim 2011) (Fig. 4.12).

4.5.2 Climatological and Climate-Change-Related Hazards

4.5.2.1 Droughts

During droughts, water levels drop, reducing the dilution of chemicals and nutrients, which worsens water quality. After droughts, heavy rains can wash these chemicals into rivers, making water even more polluted. Hot weather and anticyclones (high-pressure weather systems that block rain) increase evaporation, drying out soil and plants. This raises the chances of wildfires and reduces the soil's ability to absorb water, causing flooding when rain finally comes (Whitworth et al. 2012). Droughts also increase the risk of water contamination from animal and human waste. As water sources shrink, germs (bacteria and viruses) concentrate, raising the risk of water-borne diseases like amoebiasis (a stomach infection caused by a parasite), hepatitis A (a liver infection from a virus), salmonella (a bacterial infection causing diarrhea), schistosomiasis (a disease from parasites in freshwater), shigella (a bacterial infection causing severe diarrhea), typhoid, and paratyphoid fever (both bacterial infections that affect the stomach). While there aren't many direct links between drought and disease outbreaks, some studies report cases of E. coli O157 (a harmful bacteria), cholera (a bacterial infection causing severe diarrhea), and leptospirosis (a bacterial disease spread by animal urine) during droughts (Jackson et al. 1993; Effler et al. 2001; Tuaxe et al. 1988). When water is scarce, people may reduce hand washing, leading to more cases of diarrhoea (loose or watery bowel movements), and longer droughts are linked to higher rates of this illness (Burr 1978). Poor hygiene during droughts can also lead to skin infections like scabies (a skin condition caused by mites) and impetigo (a bacterial skin infection), and eye infections like conjunctivitis (an infection causing red, itchy eyes) (Thacker et al. 1980). An extreme example occurred in Brazil in 1996 when a drought caused cyanobacteria contamination (a type of bacteria that produces toxins) in a dialysis centre's water. This contaminated water caused severe nerve and liver damage in 126 patients, with 60 of them dying (Pouria et al. 1998). Droughts also increase airborne dust, which can cause breathing problems. During the 1930s Dust Bowl in the U.S., many people developed "dust pneumonia" (a lung infection caused by inhaling dust), which severely damaged their lungs (Cook et al. 2007). Drought followed by wet conditions can also spread diseases like St. Louis encephalitis (SLE), a virus spread by mosquitoes, as seen in Florida (Shaman et al. 2002). Similarly, more cases of West Nile virus (WNV) (a mosquito-borne virus that can cause brain swelling) have been reported during droughts in the U.S. Rift Valley fever virus (RVFV) (a viral disease from livestock) is more common after droughts during wet years. There are concerns that RVFV could spread to North America or Europe through air travel, posing a risk to human health (Weaver and Reisen 2010). Droughts also affect mental health, leading to emotional distress and even suicidal thoughts in affected individuals (Hossain et al. 2008) (Fig. 4.13).

4.5.2.2 Sea Level Rise

Rising sea levels, caused by climate change, are pushing seawater further inland, especially in flat coastal areas. This leads to stagnant water with limited flow, increasing health risks. The main health issues from sea level rise include mosquito-borne diseases (spread by mosquitoes), diseases from natural organisms (like bacteria or algae), and faecal-oral diseases (from contaminated water or food). These problems are most common in tropical areas (between 26° N to 20° N). Mosquito-borne viruses linked to rising seas include St. Louis Encephalitis (brain swelling), Eastern Equine Encephalitis (brain inflammation), West Nile virus, dengue fever (viral fever),

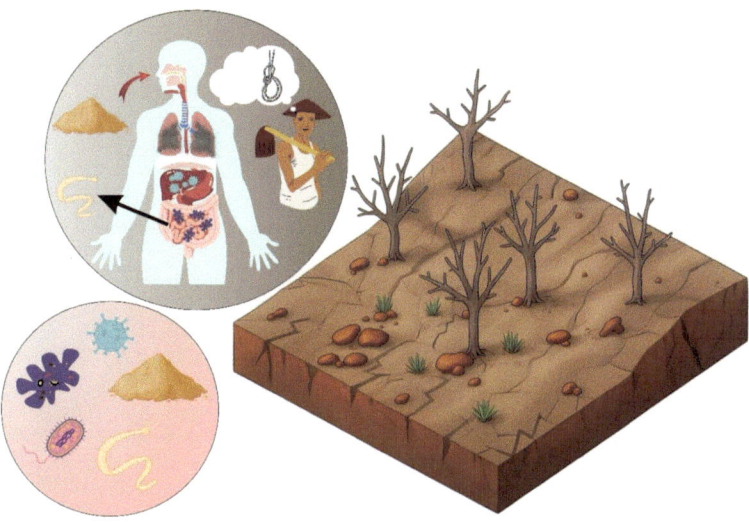

Fig. 4.13 A thematic sketch showing health hazard/destruction due to droughts. The bigger circle shows possible pathways to enter the body whereas, the smaller circle shows potential pathogens and other contaminants

Chikungunya (joint pain from a virus), Zika virus, and yellow fever (causes jaundice) (Ana et al. 2018). Diseases from natural organisms, especially bacteria like Vibrio vulnificus, Vibrio parahaemolyticus, and Vibrio cholerae, can cause vibriosis (Vibrio infections) and cholera (severe diarrhoea). People can get infected by eating raw shellfish or exposing wounds to contaminated water. Harmful algae blooms (like cyanobacteria and red tides) produce toxins that affect the digestive system (stomach and intestines), nervous system (brain and nerves), and skin, causing stomach problems and breathing issues. Avoid eating shellfish from these waters, and people with breathing problems should stay away from beaches during red tides (Carvalho et al. 2015; Newton et al. 2012). Flooding from sea level rise can damage sewers, leading to faecal-oral diseases. Common examples include infections from Campylobacter (causes diarrhoea), Giardia (a parasite causing diarrhoea), Shigella (causes dysentery), and E. coli. Protozoa-related diseases (from parasites) are especially risky, as water treatment may not kill them. Legionella bacteria, which cause Legionnaires' disease (a severe lung infection), spread through inhaling contaminated water droplets and are a concern in coastal areas (Carvalho et al. 2015; Gast et al. 2011) (Fig. 4.14).

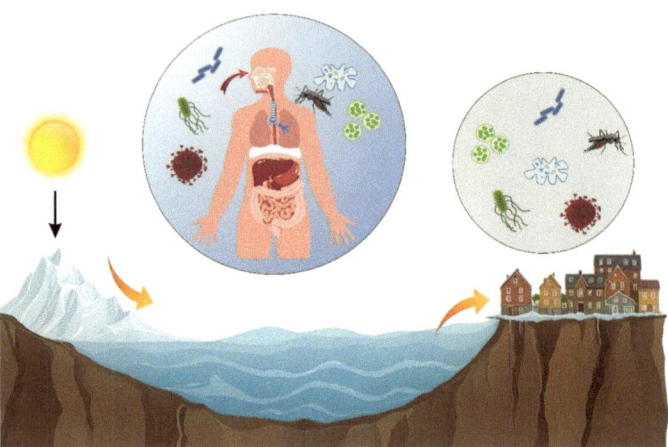

Fig. 4.14 A thematic sketch showing health hazard/destruction due to sea level rise. The bigger circle shows possible pathways to enter the body whereas, the smaller circle shows potential pathogens and other contaminants

4.5.3 *Meteorological Hazards*

4.5.3.1 Wildfires

The risk of wildfires is increasing due to climate change (Vuorio et al. 2023). As forest fires become more frequent and intense, their harmful health effects are expected to worsen (Black et al. 2017). Wildfire smoke is linked to out-of-hospital cardiac arrests (sudden heart stopping). Smoke damages the airway epithelial cells (cells lining the airways), affecting processes like autophagy (cell survival mechanism), especially in people with COPD (Chronic Obstructive Pulmonary Disease). Short-term exposure to PM2.5 (fine particles in smoke) can cause inflammation (swelling), oxidative stress (damage from oxygen imbalance), and problems in the autonomic nervous system (controls heart rate and breathing). This can lead to serious issues like vascular thrombosis (blood clots) and ventricular arrhythmias (irregular heartbeats). PM2.5 can travel from the lungs into the bloodstream, causing further body-wide inflammation. Research also links PM2.5 exposure to atherosclerotic plaques (fatty deposits in arteries), increasing the risk of heart issues (Roscioli et al. 2018; Gangwar et al. 2020). Wildfire smoke worsens cardiovascular issues (heart and blood vessel problems) like heart attacks, cardiac arrest, and heart failure, especially in those with pre-existing conditions. Smoke's fine particles can worsen endothelial dysfunction (blood vessel problems), increasing the risk of inflammation and clotting. People with diabetes, high cholesterol, or acute coronary syndrome (sudden reduced blood flow to the heart) are at higher risk (Vuorio et al. 2022; Reid et al. 2016; O'Neill et al. 2005) (Fig. 4.15).

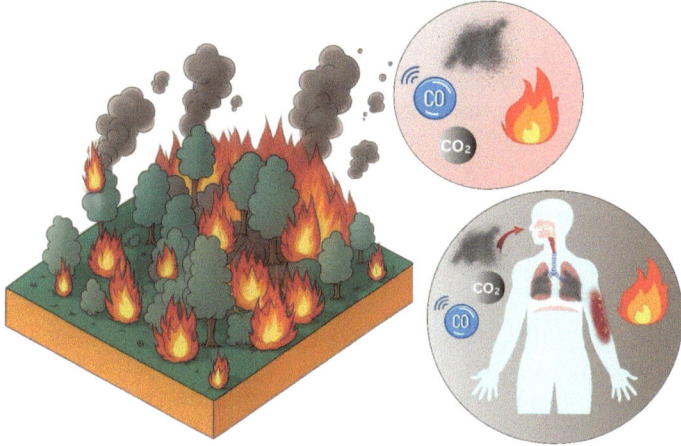

Fig. 4.15 A thematic sketch showing health hazard/destruction due to wildfires. The bigger circle shows possible pathways to enter the body whereas, the smaller circle shows potential pathogens and other contaminants

4.5.3.2 Hurricanes

Hurricanes are powerful natural disasters that can dramatically change landscapes and impact both physical and social environments. Communities often face new challenges for weeks or months after a hurricane (Waddell et al. 2021). Hurricanes can cause direct health effects, including weather-related illnesses, deaths, waterborne diseases (spread through contaminated water), vector-borne diseases (spread by insects), and mental health issues (Veenema et al. 2017). Common injuries after hurricanes include lacerations (cuts), puncture wounds, falls, and injuries from falling trees, especially during rebuilding. Drowning, poisoning, electrocution, and increased cases of vector-borne and waterborne diseases also occur. Mental health suffers too, and first responders (emergency workers) often face animal or insect bites, orthopedic injuries (injuries to bones and muscles), and cardiovascular problems (heart and blood vessel issues) (Sheppa et al. 1993; Brunkard et al. 2008). Respiratory illnesses (lung and breathing issues), such as sinusitis (inflammation of the sinus cavities) and asthma, often rise after hurricanes. Mold exposure (airborne spores from damaged homes) particularly affects children and workers. After Hurricane Katrina, some symptoms lasted months, though children's asthma improved after a year (Mitchell et al. 2012; Simeon et al. 1993). Chronic diseases like diabetes and renal failure (kidney dysfunction) can worsen after hurricanes due to medical service disruptions. There are also more pregnancy complications, including preterm deliveries (births before 37 weeks) and foetal mortality (death of a foetus). Access to contraception and HIV testing may be limited, increasing the risk of sexually transmitted diseases (STDs). Mental health impacts are significant, with people suffering from post-traumatic stress disorder (PTSD), depression, and anxiety. Substance abuse and intimate partner violence (violence between partners) tend to rise (Thethi et al. 2010; Antipova and Curtis, 2015; Leyser-Whalen et al. 2011). Hurricanes also raise the risk of infectious diseases due to overcrowded shelters and floodwaters, which can lead to skin infections, gastrointestinal disorders (stomach and intestinal issues), and mosquito-borne diseases like Zika (Grabich et al. 2004; Ferré et al. 2019; Kim et al. 2008; Harville et al. 2010; Liang et al. 2018).

4.5.3.3   Cyclones, Tornadoes and Typhoons

Cyclones can significantly impact health, especially for those with chronic conditions like diabetes (high blood sugar) and renal failure (kidney dysfunction), as medical services may be disrupted during and after the event (Cruz-Cano et al. 2019; Dosa et al. 2010; Dresser et al. 2016; Kim et al. 2017). Pregnancy complications, such as preterm birth (birth before 37 weeks) and reduced access to contraception (methods to prevent pregnancy), as well as HIV (human immunodeficiency virus) testing, may also rise (Dresser et al. 2016). Mental health is heavily impacted, with conditions like post-traumatic stress disorder (PTSD) (a mental health disorder that can occur after experiencing or witnessing a traumatic event) and depression becoming more common. Substance abuse (misusing alcohol or drugs) and intimate partner violence (violent behaviors in relationships) often increase in the aftermath (Cruz-Cano et al. 2019; Dresser et al. 2016). In overcrowded shelters, diseases spread more easily, raising the risk of skin infections (such as cellulitis), gastrointestinal disorders (problems in the stomach and intestines), respiratory illnesses (conditions affecting the lungs), and vector-borne diseases (diseases spread by insects like mosquitoes) (Parks et al. 2022). Cyclones also lead to higher death rates, particularly among older adults, and contribute to cardiovascular diseases (conditions affecting the heart and blood vessels), breathing problems, and injuries. Cyclones can worsen pregnancy outcomes, increasing the risk of low birth weight (babies born weighing less than 5.5 pounds) and other complications. Mental health issues like PTSD, anxiety, and depression become more prevalent, while infections and injuries, including traumatic brain injuries (TBI), also rise, along with opioid abuse (misusing pain-relieving drugs) and gastrointestinal illnesses (problems in the stomach and intestines) (Huang et al. 2023). Moreover, storms and heavy rain affect allergic conditions such as asthma (a chronic lung condition), allergic rhinitis (inflammation of the nasal passages), and atopic dermatitis (an itchy skin condition). Studies have shown that Mold exposure, triggered by flooding and damp conditions, can worsen allergic symptoms (Anderson et al. 2001; Rao et al. 2007). Surprisingly, some studies show a decrease in asthma cases post-storm, although this is still under investigation. Mold is a major factor in allergic rhinitis and dermatitis, with symptoms worsening in the days following heavy rain (Park et al. 2013). Tornadoes, like other natural disasters, cause significant mental health impacts. PTSD, anxiety, depression, suicidal thoughts, and substance abuse increase. Social factors like gender, age, race, financial status, and prior trauma play a role in how severely individuals are affected, with women, children, and teenagers particularly vulnerable to mental health problems (Fig. 4.16).

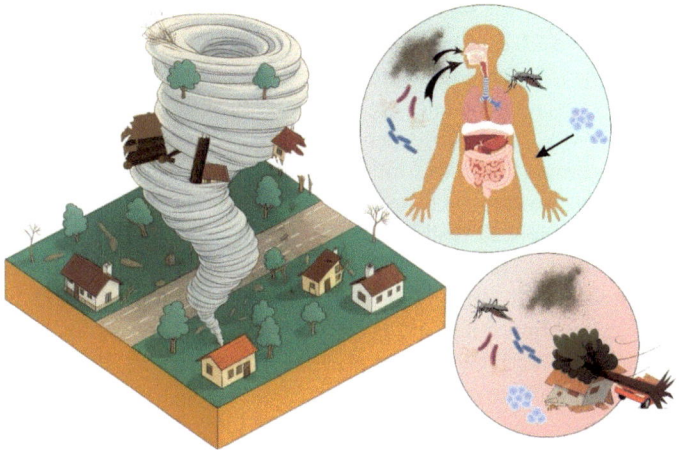

Fig. 4.16 A thematic sketch showing health hazard/destruction due to tornado. The bigger circle shows possible pathways to enter the body whereas, the smaller circle shows potential pathogens and other contaminants

4.5.3.4   Thunderstorms

During thunderstorm-associated asthma (TA) outbreaks, visits to emergency departments increase due to asthma attacks, especially in late spring and summer when pollen (tiny grains from plants) levels are high (D'Amato et al. 2016, 2017). Rain can break apart pollen grains, releasing tiny allergenic particles (substances that can cause allergies) into the air. These particles can enter deep into the lungs, causing severe reactions in people who are allergic to pollen (D'Amato and Akdis 2020). Thunderstorm winds can raise pollen levels, making asthma symptoms worse. TA events have been reported in cities like London, Melbourne, and Naples. These events are linked to thunderstorms and can affect both indoor and outdoor areas, even people who have never had asthma before. Although TA attacks don't happen all the time, their possible severity and connection to climate change show how important it is to be aware and take preventive actions (D'Amato et al. 2021). Identifying people at risk, such as those with allergic rhinitis (nasal allergies) or sensitivity to allergens, can help reduce the effects of TA outbreaks. Adding general medical practices to emergency response plans can also help manage the increased number of patients during these events (Harun et al. 2019).

4.5.3.5   Lightening

Lightning strikes cause about 24,000 deaths and many injuries worldwide each year (Sleiwah et al. 2018; Williams et al. 2018; Anketell et al. 2018). These events are sudden and can lead to serious injuries or even death. Lightning has enormous energy, with voltages over 10 million volts and currents usually between 30,000 and 110,000 amperes (Sleiwah et al. 2018; Gentges et al. 2018). People can be injured by lightning in several ways. First, the electric current can damage body tissues (cells that make up the organs and muscles). Second, the electricity turning into heat can cause burns, which are injuries to the skin and deeper tissues. Lastly, the force of the lightning can cause mechanical trauma (injuries from physical impact). Direct lightning strikes are rare but can cause severe damage by creating an electrical connection with the person (Gentges et al. 2018; Sleiwah et al. 2018; Anketell et al. 2018). Another way people are injured is called side splash, where the electric current jumps from a nearby object to a person (Sleiwah et al. 2018). The most common injury is from ground current, which happens when lightning strikes nearby and the current travels through the ground to reach the person (Gentges et al. 2018; Sleiwah et al. 2018; Anketell et al. 2018). Lightning can affect the heart, brain, and skin. A person may suddenly die from a lightning strike because both the heart and lungs may stop working at once. This is called asystolic arrest (when the heart stops beating properly), leading to death (Sleiwah et al. 2018). Lightning can also cause intracranial haemorrhage (bleeding in the brain), nerve damage, and various types of burns (Sleiwah et al. 2018; Gentges et al. 2018; Anketell et al. 2018). Treatment for lightning strike injuries involves doctors from different specialties, like emergency medicine, surgery, and cardiology (Sleiwah et al. 2018; Anketell et al. 2018; Williams et al. 2018). The first step is stabilizing the patient and ensuring the heart and breathing are working properly. Those directly struck or who experience chest pain or breathing difficulties need close monitoring and tests such as electrocardiograms (ECGs), which measure the electrical activity of the heart, and echocardiograms, which use sound waves to check the heart's health (Sleiwah et al. 2018; Gentges et al. 2018; Williams et al. 2018) (Fig. 4.17).

4.5.3.6   Dust Storms

Sand and dust storms (SDS) are worsening due to climate change, particularly in Asia, and are now a major source of harmful air pollution (Sandstrom and Forsberg 2008). These storms not only harm the environment but also affect public health, causing respiratory diseases like pneumonia (lung infection) (Kang et al. 2012; Chen et al. 2010). SDS also spread harmful microorganisms, such as bacteria and fungi, through the air, raising the risk of infections (Griffin 2007; Azua-Bustos et al. 2019). The type of germs carried by the dust can vary depending on weather and the storm's origin (Maki et al. 2017). SDS can even increase the spread of diseases like tuberculosis (TB), a serious lung infection (Wang et al. 2016; Chen et al. 2010). Understanding the microbes in SDS is crucial for assessing their health and environmental impact (Azua-Bustos et al. 2019; Rosselli et al. 2015) (Fig. 4.18).

Fig. 4.17   A thematic sketch showing health hazard/destruction due to thunderstorm and lightening. The bigger circle shows possible pathways to enter the body whereas, the smaller circle shows potential pathogens and other contaminants

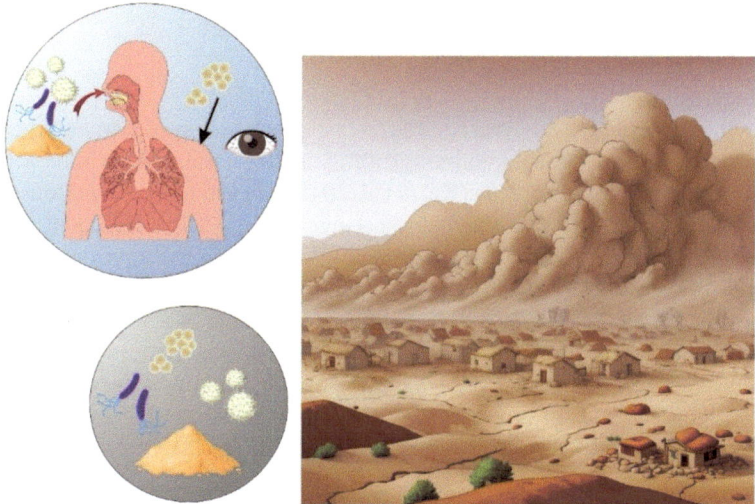

Fig. 4.18   A thematic sketch showing health hazard/destruction due to dust storm. The bigger circle shows possible pathways to enter the body whereas, the smaller circle shows potential pathogens and other contaminants

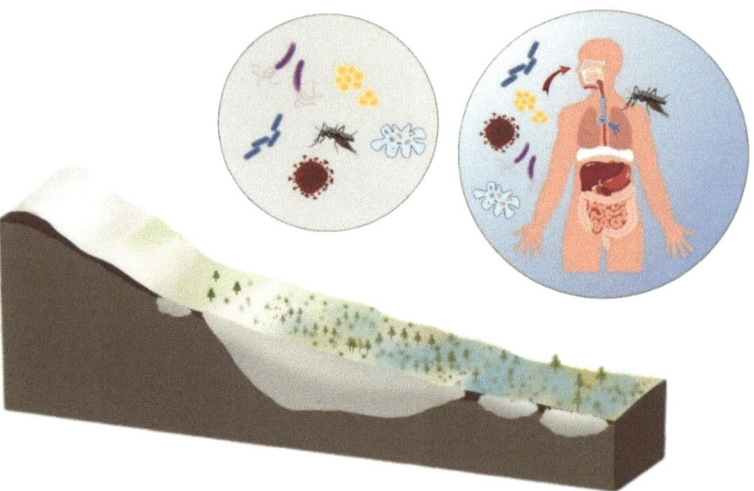

Fig. 4.19 A thematic sketch showing health hazard/destruction due to permafrost. The bigger circle shows possible pathways to enter the body whereas, the smaller circle shows potential pathogens and other contaminants

4.5.3.7   Permafrost

The thawing of permafrost in the Arctic raises concerns about the re-emergence of infectious diseases, including smallpox, anthrax, and Yersinia pestis, the bacterium responsible for the bubonic plague (Kutz et al. 2015). Historical records show that the plague caused widespread death in the past (Duggan et al. 2016). Research indicates Y. pestis can survive in permafrost, potentially re-entering the environment as the ground thaws (Petrov et al. 2018). This risk also includes viruses like the Spanish flu (H1N1), which caused the deadly 1918 pandemic (Vishnivetskaya et al. 2006). Apart from ancient diseases, melting permafrost can also release modern pollutants into local water sources, endangering the health of nearby communities (Schuur et al. 2013; Grouchy et al. 2020). Ongoing research and collaboration across fields are essential to understanding and managing these health risks (McMichael et al. 2003) (Fig. 4.19).

4.5.3.8   Hailstorm

Hailstorms form when strong winds in thunderstorms push raindrops into cold air, where they freeze into hailstones. These storms pose several health risks, especially for those outside during the storm or dealing with the aftermath. Large hailstones can cause serious injuries like concussions, head trauma, and eye damage, such as corneal abrasions (scratches on the eye) or cuts (Willmann et al. 2023; Nag 2018). Property damage can lead to emotional distress, causing anxiety and post-traumatic

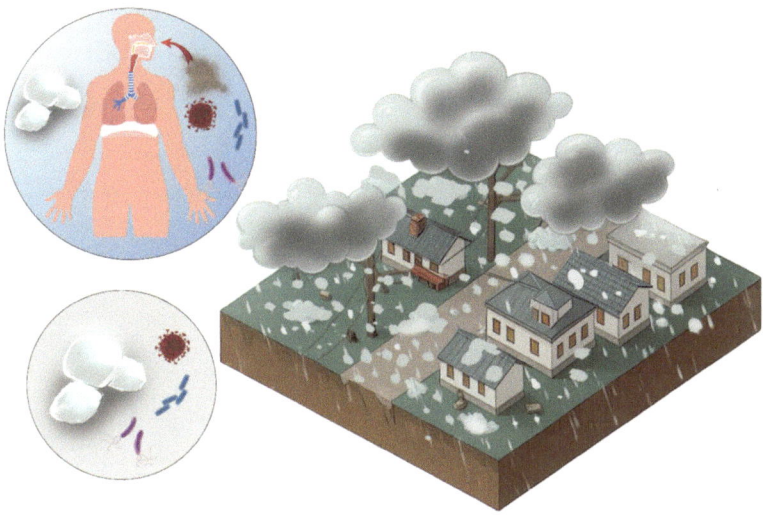

Fig. 4.20 A thematic sketch showing health hazard/destruction due to hailstorm. The bigger circle shows possible pathways to enter the body whereas, the smaller circle shows potential pathogens and other contaminants

stress disorder (PTSD) (Makwana 2019). After the storm, debris and allergens can trigger respiratory problems (Behinaein et al. 2023). Secondary hazards include contaminated water and injuries during cleanup. To reduce risks, it is important to take shelter, wear protective gear outdoors, and be cautious during post-storm cleanup (Fig. 4.20).

4.5.4 Hydrological Hazards

4.5.4.1 River Floods

Recent research highlights the strong connection between flooding and public health, especially in flood-prone areas. Floods often contaminate clean water sources with bacteria, parasites, and viruses, increasing the risk of waterborne diseases like cholera, diarrhoea, cryptosporidiosis (a parasitic infection), and rotavirus (a virus that causes severe diarrhoea) (Wakuma et al. 2019). In countries like Kenya and Vietnam, flooding is also linked to a rise in vector-borne diseases, like malaria (caused by parasites spread through mosquito bites) and Rift Valley fever (a viral disease transmitted by mosquitoes) (Ding et al. 2013). Additionally, flooding can lead to an increase in rodent-borne diseases such as leptospirosis (a bacterial infection), as rodents move into human areas following floods. These findings stress the importance of understanding the links between floods and health impacts, and implementing effective public health measures to reduce the risk to vulnerable communities (Fig. 4.21).

Fig. 4.21 A thematic sketch showing health hazard/destruction due to river floods. The bigger circle shows possible pathways to enter the body whereas, the smaller circle shows potential pathogens and other contaminants

4.5.5   Mass-Movement Hazards

4.5.5.1   Landslides

Landslides pose serious health risks, including injuries and the spread of diseases. They can contaminate water sources, leading to waterborne diseases like cholera and dysentery (Lara et al. 2009; Atuyambe et al. 2011). Stagnant water from landslides creates mosquito breeding grounds, increasing the risk of vector-borne diseases such as malaria and dengue fever (Saenz et al. 1995). Dust and particles released during landslides can also cause respiratory conditions like asthma, bronchitis, and silicosis (Redmond 2005). Infectious diseases such as leptospirosis (a bacterial infection) and hepatitis A often spread due to poor sanitation (Guha-Sapir and Below 2002; Atuyambe et al. 2011). Mental health issues, such as post-traumatic stress disorder (PTSD), depression, and anxiety, are common after landslides (Catapano et al. 2001; Tang et al. 2010). Physical injuries, including crush injuries, broken bones, and kidney failure, are also frequent and can lead to severe complications (Sever et al. 2006; Donato et al. 2011). In some cases, landslides have worsened outbreaks of diseases like malaria, as seen in Costa Rica (Saenz et al. 1995). Mental health support is critical, as many survivors experience PTSD and other issues (Norris et al. 2004; Williams et al. 2014). Addressing these risks requires quick access to clean water, medical care, and mental health services for affected communities (Fig. 4.22).

Fig. 4.22 A thematic sketch showing health hazard/destruction due to landslides. The bigger circle shows possible pathways to enter the body whereas, the smaller circle shows potential pathogens and other contaminants

4.5.5.2 Avalanches

Most avalanche deaths are due to asphyxiation, meaning the victim cannot breathe because of being buried under snow. Blunt trauma (injuries from hitting hard surfaces) causes only a small number of deaths (Falk et al. 1994; Hohlrieder et al. 2004). The cause of death can vary depending on local avalanche conditions and terrain (Stalsberg et al. 1989). Typically, victims remain alive immediately after being buried, and survival chances improve with quick rescue. However, studies show that only about 45% of users carry the recommended safety gear (Silverton et al. 2007). Air pocket devices, such as the AvaLung, can increase survival by delaying carbon dioxide buildup and maintaining oxygen levels under snow. Research shows it can keep a person oxygenated for up to 60 min (Grissom et al. 2000; Brugger et al. 2003). Fatal injuries like brain and chest trauma, however, may still occur despite rescue efforts (Grossman et al. 1989). Wearing helmets during backcountry activities is advised (Johnson et al. 2001). Although hypothermia (extreme cold exposure) can cause death, it usually occurs after asphyxiation in avalanche cases (Braun 1976; Stalsberg et al. 1989). Preventing avalanches is the best way to reduce risk, but if caught in a slide, a quick self-rescue offers the best chance for survival. Organized rescues may be too slow to be effective (Tschirky et al. 2001) (Fig. 4.23).

Fig. 4.23 A thematic sketch showing health hazard/destruction due to avalanche

4.5.5.3   Rock Fall, Debris Flow and Mudflow

Rock falls can have significant impacts on both the environment and human health. They can cause flying rocks, ground vibrations, loud noises, dust, and changes to the landscape (Turner 2018). These disturbances may lead to air and water pollution, as well as stress and anxiety in affected communities. Health issues such as respiratory problems (e.g., asthma, nasal inflammation, and colds) and hearing loss can also arise (Hazrolli 2020). Rock falls can trigger landslides, leading to property damage, economic loss, and potential loss of life (Singh et al. 2017). Mudflows from these events can damage water and sanitation systems, causing outbreaks of diseases like malaria, diarrhoea, typhoid, and cholera. For example, a cholera outbreak occurred in the region in 2012, prompting the World Health Organization (WHO) to work with local health authorities on disease prevention and response (WHO 2017). Debris flow syndrome, resulting from mudflows, involves injuries such as soft tissue damage, hypothermia, facial injuries, eye injuries (like corneal abrasions), and orthopedic injuries (Langdon 2017). Soft tissue injuries are the most common, followed by hypothermia and facial injuries. Treatment typically includes wound cleaning, surgery for soft tissue injuries, and repairing fractures and ligaments. Proper medical care can significantly reduce the severity of these injuries (Fig. 4.24).

Fig. 4.24 A thematic sketch showing health hazard/destruction due to debris flow and mud flow. The bigger circle shows possible pathways to enter the body whereas, the smaller circle shows potential pathogens and other contaminants

4.5.6 Illness Caused by Humans in Response to Environmental Changes

4.5.6.1 Disease Outbreak by Dead Bodies

Natural disaster victims usually die from trauma (injury to the body), which lowers the risk of widespread disease outbreaks. However, people handling the deceased—such as rescue workers, military personnel, and volunteers—may face infectious risks, including hepatitis B and C (liver infections), HIV (human immunodeficiency virus), enteric pathogens (bacteria or viruses that cause infections in the intestines), and tuberculosis (TB) (a bacterial infection that primarily affects the lungs). To reduce these risks, it is important to follow safety measures, such as using body bags, wearing disposable gloves, and practicing good hygiene. Vaccination against hepatitis B and TB is recommended. Burial is often the safest method for handling bodies during mass casualty events, with research showing minimal groundwater contamination (Morgon 2004). Some diseases, like TB and bloodborne viruses (hepatitis B, C, and HIV), can be transmitted from deceased bodies, especially during autopsies (post-mortem examination of the body) or if sharp instruments are involved. Though rare, diseases like Creutzfeldt-Jakob disease (a rare brain disorder) also pose risks, while meningococcal disease (a bacterial infection that causes inflammation of the membranes around the brain and spinal cord) is unlikely to spread through traditional funeral practices (WHO 2017). Group A streptococcal infections (bacterial infections that cause throat and skin infections) can lead to sore throat and skin infections, but the risk of transmission through customary practices is low for meningococcal disease.

4.5.6.2 Crowding and Diseases

Living in crowded conditions increases the risk of infectious diseases, such as tuberculosis (TB) (a bacterial infection that primarily affects the lungs), rheumatic heart disease (RHD) (a condition where the heart valves are damaged due to an infection with streptococcus bacteria), acute respiratory infections (ARIs) (infections that affect the lungs and airways), and meningococcal disease (a bacterial infection that causes serious conditions like meningitis and bloodstream infections). In overcrowded areas, germs spread more easily, leading to higher rates of these illnesses. Vulnerable groups, like those with weak immune systems, are especially at risk. For example, diseases like TB and RHD spread faster in crowded settings, and respiratory infections like the flu and respiratory syncytial virus (RSV) (a virus that causes lung infections) are more common in young children and older adults. Meningococcal disease, caused by the bacteria Neisseria meningitidis, can lead to meningitis (an infection of the membranes around the brain and spinal cord) and septicemia (blood poisoning). This disease spreads through respiratory droplets from coughing or sneezing, especially in crowded places. Improving housing conditions is crucial to help prevent the spread of these infections (Stein et al. 1950; Gray et al. 1952; Carlson et al. 1987; Baker et al. 2000; WHO 2010) (Fig. 4.25).

4.5.6.3 Bio-Terrorism

Bioterrorism is the intentional release of harmful biological agents, such as bacteria or viruses, to cause illness and disrupt society. These agents can spread through air, water, or soil, affecting large populations. In the context of medical geology, the environment plays a key role in how these agents spread. Category A agents include dangerous pathogens like anthrax and smallpox. These can survive in soil for long periods, making it easier for them to spread through the environment (Singh et al. 2017). Category B agents, such as Q fever, cause less severe illness and spread more slowly. Category C agents, like the Nipah virus and multidrug-resistant tuberculosis (MDR-TB), are emerging threats and harder to control due to their resistance to treatments (Singh et al. 2017). To prepare for bioterrorism, it's crucial to have systems in place for detecting and responding to outbreaks. This includes monitoring environmental factors that affect how these agents spread, stockpiling medical supplies like vaccines and antibiotics, and educating the public to improve readiness (Singh et al. 2017).

Fig. 4.25 Sketch showing transmission of pathogens due to crowding

4.5.6.4   Human Induced Diseases

Studies have shown that changes in the environment, driven by human activities and climate change, are closely linked to both infectious and non-infectious diseases. Climate change, which causes shifts in temperature and weather patterns, disrupts ecosystems, making it easier for vector-borne diseases, like Lyme disease (Brownstein et al. 2005) and malaria (Martens et al. 1995), to spread. These diseases are transmitted to humans through insects such as ticks and mosquitoes. Zoonotic diseases, which transfer from animals to humans, like hantavirus (Tian et al. 2017), are also increasing due to human activities, such as deforestation and urbanization. Over half of emerging infectious diseases (EIDs) are linked to changes in land use and other human behaviours (Keesing et al. 2010). Social and demographic changes, such as urbanization and global movement, contribute to the spread of diseases like HIV/AIDS (de Sousa et al. 2010) and syphilis (D'Angelo-Scott et al. 2015). Human activity has also affected wildlife populations, making it easier for diseases to spread among animals (Daszak et al. 2001). Global travel and trade have increased the spread of diseases like bubonic plague, cholera, and avian influenza (Tatem et al. 2006; Karesh et al. 2005; Naguib et al. 2015). While many viruses and bacteria are harmless, zoonotic pathogens are responsible for over 65% of new infectious disease events in the past 60 years, with 75% originating from wildlife (Keusch et al. 2009). Understanding these environmental changes and their impact on human health is crucial for addressing the rise in infectious diseases.

References

Aeroqual (2022) Respirable vs inhalable dust: what are the differences? https://www.aeroqual.com/blog/respirable-vs-inhalable-dust-differences

Al-Absi M (2013) Blue baby syndrome. Pediatric Health 8:339–404

Anderson W, Prescott GJ, Packham S, Mullins J, Brookes M, Seaton A (2001) Asthma admissions and thunderstorms: a study of pollen, fungal spores, rainfall, and ozone. QJM 94(8):429–433

Anketell J, Wilson FC, McCann J (2018) Thunder bolts and lightning': survival and neurore-habilitation following out of hospital cardiac arrest secondary to lightning strike. Brain Inj 32(12):1585–1587

Antipova A, Curtis A (2015) The post-disaster negative health legacy: pregnancy outcomes in Louisiana after Hurricane Andrew. Disasters 39:665–686. https://doi.org/10.1111/disa.12125

Askaripour M, Saeidi A, Rouleau A, Mercier-Langevin P (2022) Rockburst in underground excavations: a review of mechanism, classification, and prediction methods. Undergr Space 7(4):577–607, ISSN 2467–9674. https://doi.org/10.1016/j.undsp.2021.11.008

Atuyambe LM, Ediau M, Orach CG, Musenero M, Bazeyo W (2011) Landslide disaster in eastern Uganda: rapid assessment of water, sanitation and hygiene situation in Bulucheke camp, Bududa district. Environ Health 10:38

Azua-Bustos A, Gonzalez-Silva C, Fernandez-Martinez MA, Arenas-Fajardo C, Fonseca R, Martin-Torres FJ, Fernández-Sampedro M, Fairén AG, Zorzano M (2019) Aeolian transport of viable microbial life across the Atacama Desert, Chile: implications for mars. Sci Rep 9:11024. https://doi.org/10.1038/s41598-019-47394-z

Bajpayee TS, Rehak TR, Mowrey GL, Ingram DK (2004) Blasting injuries in surface mining with emphasis on flyrock and blast area security. J Saf Res 35(1):47–57, ISSN 0022-4375. https://doi.org/10.1016/j.jsr.2003.07.003

Baker M, McNicholas A, Garrett N et al (2000) Household crowding a major risk factor for epidemic meningococcal disease in Auckland children. Pediatr Infect Dis J 19:983–990

Barton N (2016) Cavern and tunnel collapses due to adverse structural geology (Colapsos en Túneles y Cavernas debido a Geología Estructural Adversa)

Baxter PJ, Kapila M, Mfonfu D (1989) Lake Nyos disaster, cameroon, 1986: the medical effects of large-scale emission of carbon dioxide? BMJ 298:1437–1441. An example of how good descriptive epidemiology can be carried out under difficult circumstances in remote and undeveloped areas

Baxter PJ (1990) Medical effects of volcanic eruptions. Bull Volcanol 52:532–544. One of the first articles to review health impacts of a number of volcanoes from a health perspective

Behinaein P, Hutchings H, Knapp T, Okereke IC (2023) The growing impact of air quality on lung-related illness: a narrative review. J Thorac Dis15(9):5055–5063. https://doi.org/10.21037/jtd-23-544. Epub 2023 Aug 14. PMID: 37868892; PMCID: PMC1058699

Bhandari J, Thada PK, Sedhai YR (2024) Asbestosis. [Updated 2022 Sep 19]. In: StatPearls [Internet]. Treasure Island (FL): StatPearls Publishing. Available from: https://www.ncbi.nlm.nih.gov/books/NBK555985/

Black C, Tesfaigzi Y, Bassein JA, Miller LA (2017) Wildfire smoke exposure and human health: significant gaps in research for a growing public health issue. Environ Toxicol Pharmacol 55:186–195. https://doi.org/10.1016/j.etap.2017.08.022

Bokwa A (2013) Natural hazard. In: Bobrowsky PT (eds) Encyclopedia of natural hazards. Encyclopedia of earth sciences series. Springer, Dordrecht. https://doi.org/10.1007/978-1-4020-4399-4_248

Braun P (1976) Probleme der Ersten Hilfe beim Lawinenunfall. In: Tagung Uber Medizinische Aspekte des Lawinenunfalls. Kantonsspital Zurich. Zurich, Switzerland: Juris Druck and Verlag, pp 89–96

Brownstein JS, Holford TR, Fish D (2005) Effect of climate change on Lyme disease risk in North America. EcoHealth 2:38–46. https://doi.org/10.1007/s10393-004-0139-x

Brugger H, Sumann G, Meister R et al (2003) Hypoxia and hypercapnia during respiration into an artificial air pocket in snow: implications for avalanche survival. Resuscitation 58:81–88

Brunkard J, Namulanda G, Ratard R (2008) Hurricane Katrina deaths, Louisiana, 2005. Disaster Med Pub Health Prep 2:215–223. https://doi.org/10.1097/DMP.0b013e31818aaf55

Burr ML, Davis AR, Zbijowski AG (1978) Diarrhoea and the drought. Pub Health 92(2):86–87

Carlson KH, Larsen S, Bjerve O, Leegaard J (1987) Acute bronchiolitis: predisposing factors and characterization of infants at risk. Pediatr Pulmonol 3:153–160

Carvalho DO, McKemey AR, Garziera L, Lacroix R, Donnelly CA, Alphey L et al (2015) Suppression of a field population of Aedes aegypti in Brazil by sustained release of transgenic male mosquitoes. PLoS Negl Trop Dis https://doi.org/10.1371/journal.pntd.0003864

Catapano F, Malafronte R, Lepre F, Cozzolino P, Arnone R, Lorenzo E, Tartaglia G, Starace F, Magliano L, Maj M (2001) Psychological consequences of the 1998 landslide in Sarno, Italy: a community study. Acta Psychiatr Scand 104(6):438442

Chen PS, Tsai FT, Lin CK, Yang CY, Chan CC, Young CY, Lee CH (2010) Ambient influenza and avian influenza virus during dust storm days and background days. Environ Health Perspect 118:1211–1216. https://doi.org/10.1289/ehp.0901782

CicekSenturk GAltay FA, UluKilic A, Gurbuz Y, Tutuncu E, Sencan I (2014) Acute mercury poisoning presenting as fever of unknown origin in an adult woman: a case report. J Med Case Rep

Cinelli G, De Cort M, Tollefsen T (2019) European Atlas of natural radiation. European Commission—Joint Research Centre, Luxemburg

Coalson JE, Anderson EJ, Santos EM, Madera Garcia V, Romine JK, Dominguez B, Richard DM, Little AC, Hayden MH, Ernst KC (2021) The complex epidemiological relationship between flooding events and human outbreaks of mosquito-borne diseases: a scoping review. Environ Health Perspect 129(9):96002. https://doi.org/10.1289/EHP8887. Epub 2021 Sep 28. Erratum in: Environ Health Perspect 129(12):129001. PMID: 34582261; PMCID: PMC8478154

Committee on Uranium Mining in Virginia; Committee on Earth Resources; National Research Council. Uranium Mining in Virginia: Scientific, Technical, Environmental, Human Health and Safety, and Regulatory Aspects of Uranium Mining and Processing in Virginia. Washington (DC): National Academies Press (US) (2011) Potential human health effects of uranium mining, processing, and reclamation. Available from: https://www.ncbi.nlm.nih.gov/books/NBK201047/

Cook B, Miller R, Seager R. (2007) Did dust storms make the dust bowl drought worse? Lamont-Doherty earth observatory. The Earth Institute at Colombia University. Available at: http://www.ldeo.columbia.edu/res/div/ocp/drought/dust_storms.shtml

Cruz-Cano R, Mead EL (2019) Causes of excess deaths in Puerto Rico after Hurricane Maria: a time-series estimation. Am J Pub Health 109(7):1050–1052. PMID: https://doi.org/10.2105/AJPH.2019.305015

D'Amato G, Akdis C (2020) Global warming, climate change, air pollution and allergies. Allergy 75:2158–2160

D'Angelo-Scott H, Cutler J, Friedman D, et al. Social network investigation of a syphilis outbreak in Ottawa, Ontario. Can J Infect Dis Med Microbiol 26:268–272. https://doi.org/10.1155/2015/705720

D'Amato G, Annesi Maesano I, Molino A, Vitale C, D'Amato M (2017) Thunderstorm-related asthma attacks. J Allergy Clin Immunol 139:1786–1787. [PubMed] [Google Scholar]

D'Amato G, Vitale C, D'Amato M, Cecchi L, Liccardi G, Molino A et al (2016) Thunderstorm-related asthma: what happens and why. Clin Exp Allergy 46:390–396. [PubMed] [Google Scholar]

Daszak P, Cunningham A, Hyatt A (2001) Anthropogenic environmental change and the emergence of infectious diseases in wildlife. Acta Trop 78:103–116. https://doi.org/10.1016/S0001-706 X(00)00179-0

De Sousa JD, Müller V, Lemey P, Vandamme A-M (2010) High GUD incidence in the early 20th century created a particularly permissive time window for the origin and initial spread of epidemic HIV strains. PLoS ONE 5:e9936. https://doi.org/10.1371/journal.pone.0009936

Degu Belete G, Alemu Anteneh Y (2021) General overview of radon studies in health hazard perspectives. J Oncol 2021:6659795. https://doi.org/10.1155/2021/6659795. PMID: 34381503; PMCID: PMC8352703

DenBesten P, Li W (2011) Chronic fluoride toxicity: dental fluorosis. Monogr Oral Sci 22:81–96. https://doi.org/10.1159/000327028. Epub 2011 Jun 23. PMID: 21701193; PMCID: PMC3433161

Ding G, Zhang Y, Gao L et al (2013) Quantitative analysis of burden of infectious diarrhea associated with floods in northwest of Anhui Province, China: a mixed method evaluation. PLoS One 8(6). https://doi.org/10.1371/journal.pone0065112.e65112

Donoghue AM (2004) Occupational health hazards in mining: an overview. Occup Med (Lond) 54(5):283–289. https://doi.org/10.1093/occmed/kqh072

Doocy S, Daniels A, Dick A, Kirsch TD (2013) The human impact of tsunamis: a historical review of events 1900–2009 and systematic literature review. PLoS Curr 5. https://doi.org/10.1371/currents.dis.40f3c5cf61110a0fef2f9a25908cd795. PMID: 23857277; PMCID: PMC3644289

Dosa D, Feng Z, Hyer K, Brown LM, Thomas K, Mor V (2010) Effects of Hurricane Katrina on nursing facility resident mortality, hospitalization, and functional decline. Disaster Med Pub Health Prep 4(suppl 1):S28–S32. PMID: https://doi.org/10.1001/dmp.2010.11

Donato V, Noto A, Lacquaniti A, Bolignano D, Versaci A, David A, Spinelli F, Buemi M (2011) Levels of neutrophil gelatinaseassociated lipocalin in 2 patients with crush syndrome after a mudslide. Am J Crit Care 20(5):405–409.

Dou X, Zhang S, Manzoor MU, Wen X (2023) Characteristics of methane explosion and dynamic response of rock mass in an H-type roadway with different ignition sources. ACS Omega 8(49):46513–46522. https://doi.org/10.1021/acsomega.3c04969. PMID: 38107950; PMCID: PMC10720025

Dresser C, Allison J, Broach J, Smith ME, Milsten A (2016) High-amplitude Atlantic hurricanes produce disparate mortality in small, low-income countries. Disaster Med Pub Health Prep 10(6):832–837. PMID: https://doi.org/10.1017/dmp.2016.62

Duggan AT, Perdomo MF, PiombinoMascali D, Marciniak S, Poinar D, Emery MV, Klunk J (2016) 17th century variola virus reveals the recent history of smallpox. Curr Biol 26(24):34073412

Duzgun S, Einstein H (2004) Assessment and management of roof fall risks in underground coal mines. Saf Sci 42:23–41. https://doi.org/10.1016/S0925-7535(02)00067-X

Dvorak AC, Solo-Gabriele HM, Galletti A, Benzecry B, Malone H, Boguszewski V, Bird J (2018) Possible impacts of sea level rise on disease transmission and potential adaptation strategies, a review. J Environ Manage 217:951–968, ISSN 0301-4797. https://doi.org/10.1016/j.jenvman.2018.03.102

Earle S (2019) Phys Geol. https://opentextbc.ca/geology/

Effler E, Isaacson M, Arntzen L, Heenan R, Canter P, Barrett T, Lee L, Mambo C, Levine W, Zaidi A, Griffin PM (2001) Factors contributing to the emergence of Escherichia coli O157 in Africa. Emerg Infect Dis 7(5):812–819

Falk M, Brugger H, Adler-Kastner L (1994) Avalanche survival chances. Nature 368:21

Ferré IM, Negrón S, Shultz JM, Schwartz SJ, Kossin JP, Pantin H (2019) Hurricane Maria's impact on Punta Santiago, Puerto Rico: community needs and mental health assessment six months postimpact. Disaster Med Pub Health Prep 13:18–23. https://doi.org/10.1017/dmp.2018.103

Gangwar RS, Bevan GH, Palanivel R, Das L, Rajagopalan S (2020) Oxidative stress pathways of air pollution mediated toxicity: recent insights. Redox Biol 34:101545. https://doi.org/10.1016/j.redox.2020.101545

Gast R, Moran D, Dennett M, Wurtsbaugh W, Amaral-Zettler L (2011) Amoebae and legionella pneumophila in saline environments. Water Health 9(1):37e52. https://doi.org/10.2166/wh.2010.103

Gentges J, Schieche C, Nusbaum J, Gupta N (2018) Points & Pearls: Electrical injuries in the emergency department: an evidence-based review. Emerg Med Pract 20(Suppl 11):1–2

Geohazards (2019) Alaska science and nature. https://www.nps.gov/subjects/aknatureandscience/geohazards.htm

Grabich SC, Robinson WR, Konrad CE, Horney JA (2017) Impact of Hurricane exposure on reproductive health outcomes, Florida, 2004. Disaster Med Pub Health Prep 11:407–411. https://doi.org/10.1017/dmp.2016.158

Graczyk H, Azzi M, Mandrioli D (2021) International labour organization—exposure to hazardous chemicals at work and resulting health impacts: a global review

Gray FG, Quinn RW, Quinn JP (1952) A long-term survey of rheumatic and non-rheumatic families; with reference to environment and heredity. Am J Med 13:400–412

Griffin DW (2007) Atmospheric movement of microorganisms in clouds of desert dust and implications for human health. Clin Microbiol Rev 20:459–477. https://doi.org/10.1128/CMR.00039-06

Grissom CK, Radwin MI, Harmston CH et al (2000) Respiration during snow burial using an artificial air pocket. JAMA 283:2266–2271

Grouchy M, López-Merino L, Allard R, Christensen TR, Legrand D, Normand M, Al T (2020) Experimental study of thawing permafrost microbial communities in a unique Arctic lake: greenhouse gas dynamics and methanogenic microorganisms. Environ Microbiol 22(3):10861102

Grossman MD, Saffle JR, Thomas F et al (1989) Avalanche trauma. J Trauma 29:1705–1709

GuhaSapir D, Below R (2002) The quality and accuracy of disaster data, a comparative analyses of three global data sets, a working paper prepared for the Disaster Management Facility, World Bank. WHO Centre for Research on the Epidemiology of Disasters, University of Louvain School of Medicine, Brussels

Han S, Chen H, Harvey MA, Stemn E, Cliff D (2018) Focusing on coal workers' lung diseases: a comparative analysis of China, Australia, and the United States. Int J Environ Res Pub Health 15(11):2565. https://doi.org/10.3390/ijerph15112565. PMID: 30453500; PMCID: PMC6266950

Hansell A, Oppenheimer C (2004) Health hazards from volcanic gases: a systematic literature review. Arch Environ Health 59(12):628–639. https://doi.org/10.1080/00039890409602947. PMID: 16789471

Hansell AL, Horwell CJ, Oppenheimer C (2006) The health hazards of volcanoes and geothermal areas. Occup Environ Med 63(2):149–156, 125. https://doi.org/10.1136/oem.2005.022459. PMID: 16421396; PMCID: PMC2078062

Harun NS, Lachapelle P, Douglass J (2019) Thunderstorm-triggered asthma: what we know so far. J Asthma Allergy 12:101–108

Harville EW, Taylor CA, Tesfai H, Xiong X, Buekens P (2010) Experience of Hurricane Katrina and reported intimate partner violence. J Interpers Violence 26:833–845. https://doi.org/10.1177/0886260510365861

Hasan SE (2021) Medical geology. In: Encyclopedia of geology, vol. 684. Cambridge: Academic

Hazrolli V (2020) Gurethyesit dhendikimii tyre ne mjedis

Hohlrieder M, Eschertzhuber S, Schubert H et al (2004) Severity and pattern of injury in survivors of alpine fall accidents. High Alt Med Biol 5:349–354

Hossain D, Eley R, Coutts J, Gorman D (2008) The mental health of landholders in Southern Queensland: issues and support. Aust J Rural Health 16(6):343–348

Huang W, Gao Y, Xu R, Yang Z, Yu P, Ye T, Ritchie EA, Li S, Guo Y (2023) Health effects of cyclones: a systematic review and meta-analysis of epidemiological studies. Environ

Health Perspect 131(8):86001. https://doi.org/10.1289/EHP12158. Epub 2023 Aug 28. PMID: 37639476; PMCID: PMC10461789

Jackson LA, Kaufmann AF, Adams WG, Phelps MB, Andreasen C, Langkop CW, Francis BJ, Wenger JD (1993) Outbreak of leptospirosis associated with swimming. Pediatr Infect Dis J 12(1):48–54

Jiménez-Forero CP, Zabala II, Idrovo AJ (2015) Work conditions and morbidity among coal miners in Guachetá, Colombia: the miners´ perspective. Biomédica 35:77–89. https://doi.org/10.7705/biomedica.v35i0.2439

Johnson SM, Johnson AC, Barton RG (2001) Avalanche trauma and closed head injury: adding insult to injury. Wilderness Environ Med 12:244–247

Kang JH, Keller JJ, Chen CS, Lin HC (2012) Asian dust storm events are associated with an acute increase in pneumonia hospitalization. Ann Epidemiol 22:257–263. https://doi.org/10.1016/j.annepidem.2012.02.008

Keesing F, Belden LK, Daszak P et al (2010) Impacts of biodiversity on the emergence and transmission of infectious diseases. Nature 468:647–652. https://doi.org/10.1038/nature09575

Keim M (2011) The public health impact of tsunami disasters. Am J Disaster Med 6:341–349. https://doi.org/10.5055/ajdm.2011.0073

Keusch GT, Pappaioanou M, Gonzalez MC et al (2009) Sustaining global surveillance and response to emerging zoonotic diseases. National Academic Press, New York

Kim SC, Plumb R, Gredig QN, Rankin L, Taylor B (200) Medium-term post-Katrina health sequelae among New Orleans residents: predictors of poor mental and physical health. J Clin Nurs 17:2335–2342. https://doi.org/10.1111/j.1365-2702.2008.02317.x

Kim S, Kulkarni PA, Rajan M, Thomas P, Tsai S, Tan C et al (2017) Hurricane Sandy (New Jersey): mortality rates in the following month and quarter. Am J Pub Health 107(8):1304–1307. PMID: https://doi.org/10.2105/AJPH.2017.303826

Kistnasamy B, Yassi A, Yu J et al (2018) Tackling injustices of occupational lung disease acquired in South African mines: recent developments and ongoing challenges. Glob Health 14:60. https://doi.org/10.1186/s12992-018-0376-3

Knobeloch L, Salna B, Hogan A, Postle J, Anderson H (2000) Blue babies and nitrate-contaminated well water. Environ Health Perspect 108(7):675–678. https://doi.org/10.1289/ehp.00108675. PMID: 10903623; PMCID: PMC1638204

Komac B, Zorn M (2013) Geohazards. In: Bobrowsky PT (eds) Encyclopedia of natural hazards. Encyclopedia of earth sciences series. Springer, Dordrecht. https://doi.org/10.1007/978-1-4020-4399-4_154

Kulkarni NP, Mandal BB (2015) An evaluative study of occupational noise exposure for operators in bauxite mines. In: National conference on safety & health management systems to improve productivity in mines, Nagpur, pp 23–25

Kutz SJ, Hoberg EP, Polley L, Jenkins EJ (2015) Global warming is changing the dynamics of Arctic host-parasite systems. Proc Roy Soc B: Biol Sci 282(1821):20150351

Kyeremateng-Amoah E, Clarke EE (2015) Injuries among artisanal and small-scale gold miners in Ghana. Int J Environ Res Pub Health 12(9):10886–10896. https://doi.org/10.3390/ijerph120910886

Langdon S, Johnson A, Sharma R (2019) Debris flow syndrome: injuries and outcomes after the Montecito debris flow. Am Surg 85(10):1094–1098. PMID: 31657301

Lara RJ, Neogi SB, Islam MS, Mahmud ZH, Yamasaki S, Nair GB (2009) Influence of catastrophic climatic events and human waste on Vibrio distribution in the Karnaphuli estuary. Bangladesh. Ecohealth 6(2):279286

Leyser-Whalen O, Rahman M, Berenson AB (2011) Natural and social disasters: racial inequality in access to contraceptives after Hurricane Ike. J Women's Health 20:1861–1866

Liang SY, Messenger N (2018) Infectious diseases after hydrologic disasters. Emerg Med Clin N Am 36:835–851. https://doi.org/10.1016/j.emc.2018.07.002

Lin S, Liu Z, Qian J, Li X, Zhang Q (2019) Flammability and explosion risk of post-explosion CH 4/air and CH 4/coal dust/air mixtures. Combust Sci Technol 193:1–14. https://doi.org/10.1080/00102202.2019.1688313

Lucchini RG, Martin CJ, Doney BC (2009) From manganism to manganese-induced parkinsonism: a conceptual model based on the evolution of exposure. Neuromol Med 11(4):311–321. https://doi.org/10.1007/s12017-009-8108-8. Epub 2009 Dec 10. PMID: 2001238

Maier RM, Díaz-Barriga F, Field JA, Hopkins J, Klein B, Poulton MM (2014) Socially responsible mining: the relationship between mining and poverty, human health and the environment. Rev Environ Health 29:83–89. https://doi.org/10.1515/reveh-2014-0022

Maki T, Kurosaki Y, Onishi K, Lee KC, Pointing SB, Jugder D, Yamanaka N, Hasegawa H, Shinoda M (2017) Variations in the structure of airborne bacterial communities in Tsogt-Ovoo of Gobi desert area during dust events. Air Qual Atmos Health 10:249–260. https://doi.org/10.1007/s11869-016-0430-3

Makwana N (2019) Disaster and its impact on mental health: a narrative review. J Family Med Prim Care. 8(10):3090–3095. https://doi.org/10.4103/jfmpc.jfmpc_893_19.PMID:31742125; PMCID:PMC6857396

Mandal B, Srivastava A (2006) Risk from vibration in Indian mines. Ind J Occup Environ Med 10. https://doi.org/10.4103/0019-5278.27460

Manwar VD, Mandal BB, Pal AK (2016) Environmental propagation of noise in mines and nearby villages: a study through noise mapping. Noise Health 18(83):185–193. https://doi.org/10.4103/1463-1741.189246. PMID: 27569406; PMCID: PMC5187660

Martens W, Niessen LW, Rotmans J et al (1995) Potential impact of global climate change on malaria risk. Environ Health Perspect 103:458. https://doi.org/10.1289/ehp.95103458

Mastrolorenzo G, Petrone PP, Pagano M et al (2001) Herculaneum victims of Vesuvius in AD79. Nature 410:769–770

Matamala Pizarro J, Aguayo Fuenzalida F (2021) Mental health in mine workers: a literature review. Ind Health 59(6):343–370. https://doi.org/10.2486/indhealth.2020-0178. Epub 2021 Sep 28. PMID: 34588377; PMCID: PMC8655752

Mavrouli M, Mavroulis S, Lekkas E, Tsakris A (2023) The impact of earthquakes on public health: a narrative review of infectious diseases in the post-disaster period aiming to disaster risk reduction. Microorganisms 11(2):419. https://doi.org/10.3390/microorganisms11020419. PMID: 36838384; PMCID: PMC9968131

Mayer A, Hamzeh N (2015) Beryllium and other metal-induced lung disease. Curr Opin Pulm Med 21(2):178–184

McMichael AJ, Woodruff RE, Hales S (2003) Climate change and human health: present and future risks. The Lancet 367(9513):859869

Mitchell H, Cohn RD, Wildfire J, Thornton E, Kennedy S, El-Dahr JM, Chulada PC, Mvula MM, Grimsley LF, Lichtveld MY et al (2012) Implementation of evidence-based Asthma Interventions in Post-Katrina New Orleans: the Head-off Environmental Asthma in Louisiana (HEAL) Study Environ Health Perspect 120:1607–1612. https://doi.org/10.1289/ehp.1104242

Mines Safety and Inspection Act (1994) Department of energy, mines, industry regulation and safety. Minister for Mines and Petroleum https://www.dmp.wa.gov.au/Safety/Guidance-about-fire-hazards-6383.aspx

Morgon O (2004) Infectious disease risks from dead bodies following natural disasters. Rev Panam Salud Publica 15(5):307–312. https://doi.org/10.1590/s1020-49892004000500004. PMID: 15231077

Mueller W, Cowie H, Horwell CJ, Hurley F, Baxter PJ (2020) Health impact assessment of volcanic ash inhalation: a comparison with outdoor air pollution methods. Geohealth 4(7):e2020GH000256. https://doi.org/10.1029/2020GH000256. PMID: 32642627; PMCID: PMC7334379

Nag OS (2018) Environment home environment what are the dangerous effects of a hailstorm? https://www.worldatlas.com/articles/what-are-the-dangerous-effects-of-a-hailstorm.html

Naguib MM, Kinne J, Chen H et al (2015) Outbreaks of highly pathogenic avian influenza H5N1 clade 2.3. 2.1 c in hunting falcons and kept wild birds in Dubai implicate intercontinental virus spread. J Gen Virol 96:3212–3222. https://doi.org/10.1099/jgv.0.000274.

National Cancer Institute (2022) Asbestos. https://www.cancer.gov/about-cancer/causes-preven tion/risk/substances/asbestos

Newton A, Kendall M, Vugia DJ, Henao OL, Mahon BE (2012) Increasing rates of vibriosis in the United States, 1996–2010: review of surveillance data from 2 systems. Clin Infect Dis 54(suppl_5):S391–S395

Noji E (1997) Earthquakes. In: Noji E (ed) Public health consequences of disasters. Oxford University Press, New York, NY

Norris FH, Murphy AD, Baker CK, Perilla JL (2004) Postdisaster PTSD over four waves of a panel study of Mexico's 1999 flood. J Trauma Stress 17(4):283292

Omidi L, Zare S, Rad R, Meshkani M, Kalantari S (2017) Effects of shift work on health and satisfaction of workers in the mining industry. Int J Occup Hyg 9

Onder S, Mutlu M (2016) Analyses of non-fatal accidents in an opencast mine by logistic regression model—a case study. Int J Inj Control Saf Promot 24(3):328–337. https://doi.org/10.1080/174 57300.2016.1178299

O'Neill MS, Veves A, Zanobetti A, Sarnat JA, Gold DR, Economides PA et al (2005) Diabetes enhances vulnerability to particulate air pollution-associated impairment in vascular reactivity and endothelial function. Circulation 111:2913–2920. https://doi.org/10.1161/CIRCUL ATIONAHA.104.517110

Park KJ, Moon JY, Ha JS, Kim SD, Pyun BY, Min TK, Park YH (2013) Impacts of heavy rain and typhoon on allergic disease. Osong Pub Health Res Perspect 4(3):140–145. https://doi.org/10. 1016/j.phrp.2013.04.009. Epub 2013 Apr 30. PMID: 24159545; PMCID: PMC3787533

Parks RM, Benavides J, Anderson GB, Nethery RC, Navas-Acien A, Dominici F et al (2022) Association of tropical cyclones with county-level mortality in the US. JAMA 327(10):946–955. PMID: https://doi.org/10.1001/jama.2022.1682

Pelucchi C, Pira E, Piolatto G, Coggiola M, Carta P, La Vecchia C (2006) Occupational silica exposure and lung cancer risk: a review of epidemiological studies 1996-2005. Ann Oncol 17(7):1039–50. https://doi.org/10.1093/annonc/mdj125. Epub 2006 Jan 10. PMID: 16403810

Petrov DA, Loeschcke V, Zhivotovsky LA (2018) The emergence of novel diseases in frozen environments. Eur J Hum Genet 26(5):547549

Pouria S, de Andrade A, Barossa J, Cavalcanti R, Barreto V, Ward C, Preiser W, Poon G, Neild G, Codd G (1998) Fatal microcystin intoxication in haemodialysis unit in Caruaru, Brazil. Lancet 352:21–26

Ramírez-Castillo FY, Loera-Muro A, Jacques M, Garneau P, Avelar-González FJ, Harel J, Guerrero-Barrera AL (2015) Waterborne pathogens: detection methods and challenges. Pathogens 4(2):307–334. https://doi.org/10.3390/pathogens4020307. PMID: 26011827; PMCID: PMC4493476

Randive K (2013) Elements of geochemistry. Geochem Explor Med Geol

Rao CY, Riggs MA, Chew GL (2007) Characterization of airborne molds, endotoxins, and glucans in homes in New Orleans after Hurricanes Katrina and Rita. Appl Environ Microbiol 73(5):1630–1634

Redmond AD (2005) ABC of conflict and disaster: natural disasters. BMJ 330(7502):12591261

Reid CE, Brauer M, Johnston FH, Jerrett M, Balmes JR, Elliot CT (2016) Critical review of health impacts of wildfire smoke exposure. Environ Health Perspect 124:1334–1343. https://doi.org/ 10.1289/ehp.1409277

Requena-Mullor M, Alarcón-Rodríguez R, Parrón-Carreño T, Martínez-López JJ, Lozano-Paniagua D, Hernández AF (2021) Association between crystalline silica dust exposure and silicosis development in artificial stone workers. Int J Environ Res Pub Health. 18(11):5625. https://doi. org/10.3390/ijerph18115625. PMID: 34070293; PMCID: PMC8197517

Resource safety and Health Queensland (2023) Managing heat exposure in coal mines. Mines Saf Bull 191. Version 3 https://www.rshq.qld.gov.au/

Riudavets M, Garcia de Herreros M, Besse B, Mezquita L (2022) Radon and lung cancer: current trends and future perspectives. Cancers (Basel) 14(13):3142. https://doi.org/10.3390/cancers14 133142. PMID: 35804914; PMCID: PMC9264880

Roscioli E, Hamon R, Lester SE, Jersmann HPA, Reynolds PN, Hodge S (2018) Airway epithelial cells exposed to wildfire smoke extract exhibit dysregulated autophagy and barrier dysfunction consistent with COPD. Respir Res 19:234. https://doi.org/10.1186/s12931-018-0945-2

Rowens B, GuerreroBetancourt D, Gottlieb CA, Boyes RJ, Eichenhorn MS (1991) Respiratory failure and death following acute inhalation of mercury vapor. Chest

Roy S, Mishra DP, Bhattacharjee RM et al (2022) Heat stress in underground mines and its control measures: a systematic literature review and retrospective analysis. Min Metall Explor 39:357–383. https://doi.org/10.1007/s42461-021-00532-6

Rumchev K, Van Hoang D, Lee AH (2023) Exposure to dust and respiratory health among Australian miners. Int Arch Occup Environ Health 96(3):355–363. https://doi.org/10.1007/s00420-022-01922-z. Epub 2022 Sep 12. Erratum in: Int Arch Occup Environ Health 2023 Feb 1: PMID: 36089622; PMCID: PMC9968258

Saenz R, Bissell RA, Paniagua F (1995) Postdisaster malaria in Costa Rica. Prehospital Disaster Med 10(3):154160

Salcioglu E, Basoglu M, Livanou M (2007) Post-traumatic stress disorder and comorbid depression among survivors of the 1999 earthquake in Turkey. Disasters 31:115–129. https://doi.org/10.1111/j.1467-7717.2007.01000.x

Salve UR, Paul A (2022) Ergonomics in mining: current status and future challenges. In: Randive K, Pingle S, Agnihotri A (eds) Medical geology in mining. Springer Geology. Springer, Cham. https://doi.org/10.1007/978-3-030-99495-2_11

Sandstrom T, Forsberg B (2008) Desert dust: An unrecognized source of dangerous air pollution? Epidemiology 19:808–809. https://doi.org/10.1097/EDE.0b013e31818809e0

Schneider E, Hajjeh RA, Spiegel RA, Jibson RW, Harp EL, Marshall GA, Gunn RA, McNeil MM, Pinner RW, Baron RC et al (1997) A coccidioidomycosis outbreak following the Northridge, Calif, earthquake. J Am Med Assoc JAMA 277:904–908. https://doi.org/10.1001/jama.1997.03540350054033

Schuur EAG, McGuire AD, Schädel C, Grosse G, Harden JW, Hayes DJ, Hugelius G (2013) Climate change and the permafrost carbon feedback. Nature 520(7546):171179

Sever MS, Vanholder R, Lameire N (2006) Management of crush related injuries following disasters. N Engl J Med 354:10521063

Shaman J, Day J, Stieglitz M (2002) Drought-induced amplification of Saint Louis encephalitis virus, Florida. Emerg Infect Dis 8(6):575–580

Shekarian Y, Rahimi E, Rezaee M, Su WC, Roghanchi P (2021) Respirable coal mine dust: a review of respiratory deposition, regulations, and characterization. Minerals 11(7):696. https://doi.org/10.3390/min11070696

Sheppa C, Stevens J, Philbrick JT, Canada M (1993) The effect of a class IV hurricane on emergency department operations. Am J Emerg Med 11:464–467. https://doi.org/10.1016/0735-6757(93)90084-O

Silverton N, McIntosh SE, Kim H (2007) Avalanche safety practices in Utah. Wilderness Environ Med 18:264–270

Simeon DT, Grantham-McGregor SM, Walker SP, Powell CA (1993) Effects of a hurricane on growth and morbidity in children from low-income families in Kingston, Jamaica. Trans R Soc Trop Med Hyg 87:526–528. https://doi.org/10.1016/0035-9203(93)90073-Y

Singh A et al (2017) Impact of landslides on environment; Selgelid MJ (2012) Bioterrorism. In: Encyclopedia of applied ethics, 2nd edn. Academic Press, pp 309–316. ISBN 9780123739322. https://doi.org/10.1016/B978-0-12-373932-2.00025-9

Sizar O, Berylliosis TR (2024) [Updated 2023 Feb 5]. In: StatPearls [Internet]. Treasure Island (FL): StatPearls Publishing. Available from: https://www.ncbi.nlm.nih.gov/books/NBK470364/

Sleiwah A, Baker J, Gowers C, Elsom DM, Rashid A (2018) Lightning injuries in Northern Ireland. Ulster Med J 87(3):168–172

Stalsberg H, Albretsen C, Gilbert M et al (1989) Mechanism of death in avalanche victims. Virchows Arch A Pathol Anat Histopathol 414:415–422

Steen TW, Gyi KM, White NW et al (1997) Prevalence of occupational lung disease among Botswana men formerly employed in the South African mining industry. Occup Environ Med 54:19–26. https://doi.org/10.1136/oem.54.1.19

Stein L (1950) A study of respiratory tuberculosis in relation to housing conditions in Edinburgh; the pre-war period. Brit J Soc Med 4:143–169

Tatem AJ, Rogers DJ, Hay S (2006) Global transport networks and infectious disease spread. Adv Parasitol 62:293–343. https://doi.org/10.1016/S0065-308X(05)62009-X

Tauxe RV, Holmberg SD, Dodin A, Wells JV, Blake PA. (1988) Epidemic cholera in Mali: high mortality and multiple routes of transmission in a famine area. Epidemiol Infect 100(2):279–289

Tang TC, Yen CF, Cheng CP, Yang P, Chen CS, Yang RC, Huang MS, Jong YJ, Yu HS (2010) Suicide risk and its correlate in adolescents who experienced typhoon induced mudslides: a structural equation model. Depress Anxiety 27(12):1143–1148.

Thacker SB, Music SI, Pollard RA, Berggren G, Boulos C, Nagy T, Brutus M, Pamphile M, Ferdinand RO, Joseph VR (1980) Acute water shortage and health problems in Haiti. Lancet 1(8166)471–473

Thakkar L, Jain RK, Pingle S, Barde S, Arakera SB (2022) The scope for early diagnosis of noise-induced hearing loss among mine and industrial workers: a brief review. In: Randive K, Pingle S, Agnihotri A (eds) Medical geology in mining. springer geology. Springer, Cham. https://doi.org/10.1007/978-3-030-99495-2_8

Thethi TK, Yau CL, Shi L, Leger S, Nagireddy P, Waddadar J, Surampudi P, John-Kalarickal J, Yenoby L, Fonseca V (2010) Time to recovery in diabetes and comorbidities following Hurricane Katrina. Disaster Med Pub Health Prep 4:S33–S38. https://doi.org/10.1001/dmp.2010.10

Tian H, Yu P, Bjørnstad ON et al (2017) Anthropogenically driven environmental changes shift the ecological dynamics of hemorrhagic fever with renal syndrome. PLoS Pathog 13:e1006198. https://doi.org/10.1371/journal.ppat.1006198

Tschirky F, Brabec B, Kern M (2001) Avalanche rescue devices, state of the development, success and failure [in German]. In: Osterreichische Gesellschaft fur Alpin und Hohenmedizin. Austrian Society of Mountain Medicine, pp 101–125

Turner AK (2018) Social and environmental impacts of landslides. Innovative Infrastruct Solut 28(Sep):41

Veenema TG, Thornton CP, Lavin RP, Bender AK, Seal S, Corley A (2017) Climate change-related water disasters' impact on population health. J Nurs Sch 49:625–634. https://doi.org/10.1111/jnu.12328

Vishnivetskaya TA, Petrova MA, Urbance J, Ponder M, Moyer CL, Gilichinsky DA, Tiedje JM (2006) Bacterial community in ancient Siberian permafrost as characterized by culture and culture independent methods. Astrobiology 6(3):400414

Vuorio A, Budowle B, Kovanen PT (2022) Airborne particles and cardiovascular morbidity in severe inherited hypercholesterolemia: vulnerable endothelium under multiple attacks. BioEssays 44:e2100273. https://doi.org/10.1002/bies.202100273

Vuorio A, Budowle B, Raal F, Kovanen PT (2023) Wildfire smoke exposure and cardiovascular disease-should statins be recommended to prevent cardiovascular events? Front Cardiovasc Med 14(10):1259162. https://doi.org/10.3389/fcvm.2023.1259162.PMID:37781301;PMCID: PMC10537918

Waddell SL, Jayaweera DT, Mirsaeidi M, Beier JC, Kumar N (2021) Perspectives on the health effects of hurricanes: a review and challenges. Int J Environ Res Pub Health 18(5):2756. https://doi.org/10.3390/ijerph18052756. PMID: 33803162; PMCID: PMC7967478

Wakuma AS, Mandere N, Ewald G (2019) Floods and health in Gambella region, Ethiopia: a qualitative assessment of the strengths and weaknesses of coping mechanisms. Glob Health Action 2(1):2019. https://doi.org/10.3402/gha.v2i0.2019.[PMCfreearticle][PubMed][CrossRef][GoogleScholar]

Wang Y, Wang R, Ming J, Liu G, Chen T, Liu X, Liu H, Zhen Y, Cheng G (2016) Effects of dust storm events on weekly clinic visits related to pulmonary tuberculosis disease in Minqin, China. Atmos Environ 127:205–212. https://doi.org/10.1016/j.atmosenv.2015.12.041

Weaver S, Reisen W (2010) Present and future arboviral threats. Antiviral Res 85:328–345

Wesdock JC, Arnold IMF (2014) Occupational and environmental health in the aluminium industry: key points for health practitioners. J Occup Environ Med 56(5 Suppl):S5–S11. https://doi.org/10.1097/JOM.0000000000000071

Whitworth K, Baldwin D, Kerr J (2012) Drought, floods, and water quality: drivers of a severe hypoxic blackwater event in a major river system (the southern Murray-Darling Basin, Australia). J Hydrol 450–451, 190–198

WHO Sierra Leone (2017) WHO provides urgent health assistance to meet needs of people affected by floods and landslides in Sierra Leone. Retrieved from https://www.afro.who.int/news/who-provides-urgent-health-assistance-meet-needs-people-affected-floods-and-landslides-sierra

WHO (2009) Handbook on indoor radon: a public health perspective. World Health Organization; Geneva, Switzerland

Wick K, Heumesser C, Schmid E (2012) Groundwater nitrate contamination: factors and indicators. J Environ Manage 111(3):178–186. https://doi.org/10.1016/j.jenvman.2012.06.030. Epub 2012 Aug 18. PMID: 22906701; PMCID: PMC3482663

Williams R, Greenberg N (2014) Psychosocial and mental health care for the deployed staff of rescue, professional first response and aid agencies, NGOs and military organisations. in: conflict and catastrophe medicine: a practical guide, 3rd edn. Springer, London

Williams VF, Oetting AA, Stahlman S (2018) Update: lightning strike injuries, active component, U.S. Armed Forces, 2008–2017. MSMR 25(9):20–24

Willmann D, Fu L, Melanson SW. Corneal injury. [Updated 2023 Jul 17]. In: StatPearls [Internet]. Treasure Island (FL): StatPearls Publishing. Available from: https://www.ncbi.nlm.nih.gov/books/NBK459283/

Woodward A, Hales S (2014) The past and future of coal. Aust N Z J Pub Health 38(2):103–104. https://doi.org/10.1111/1753-6405.12215

World Health Organization (2010) Housing and health guidelines. World Health Organization, Geneva

Wu X, Yin W, Wu C, Li Y (2017) Development and validation of a safety attitude scale for coal miners in China. Sustainability 9(12). https://doi.org/10.3390/su9122165

Witham CS, Oppenheimer C, Horwell CJ (2005) Volcanic ash-leachates: a review and recommendations for sampling methods. J Volcanol Geotherm Res 141:299–326

Witham CS, Oppenheimer C (2004) Mortality in England during the 1783–4 Laki Craters eruption. Bull Volcanol 6715

Yorifuji T, Sato T, Yoneda T, Kishida Y, Yamamoto S, Sakai T, Sashiyama H, Takahashi S, Orui H, Kato D et al (2018) Disease and injury trends among evacuees in a shelter located at the epicenter of the 2016 Kumamoto earthquakes. Jpn Arch Environ Occup Health 73:284–291. https://doi.org/10.1080/19338244.2017.1343238

Yassi A (1997) Repetitive strain injuries. Lancet 349(9056):943–947. https://doi.org/10.1016/S0140-6736(96)07221-2. PMID: 9093264

Chapter 5
Geo-Pharmacy

This chapter discusses therapeutic practices for curing diseases by using geo-materials such as minerals, rocks, fossils, petroleum products, etc. Geo-Pharmacy is the term used for ancient medicines using geomaterials, whereas, the modern definition more elaborately includes wholistic treatments using geomaterials such balneotherapy and thalassotherapy emphasizing the healing properties of natural mineral springs and seawater, as well as skin treatments using clays (mud bath), gemstones and other crystals. Mercurial medicines used in Ayurveda depicts ancient knowledge of using geomaterials, whereas, geophagy is based on traditional medicinal knowledge inherited by tribals. Finally, cryotherapy underscores the physiological benefits of cold temperatures in promoting health and well-being.

5.1 Basic Concepts of Geo-Pharmacy

The geomaterials are the important sources of medicines that can be used for treatment of a number of health ailments. Numerous books, articles, and symposia have focused on health issues related to elements like arsenic, mercury, and lead, minerals such as asbestos and quartz, and geologic events like earthquakes and volcanic eruptions; however, the geomaterials find number of used in treatment of ailments (Limpitlaw 2004). The health benefits of geologic materials and processes have a long history, although this knowledge has diminished over time, and with the advent of modern (allopathic) medicine, the loss of indigenous knowledge has been detrimental. Evidence suggests that early humans, such as Homo habilis, used powdered clays for digestion and treating upset stomachs two million years ago. For thousands of years, ancient civilizations like those in Mesopotamia, China, India, and Egypt used minerals for their therapeutic properties (Finkelman 2006). One of the earliest known medicinal uses of geologic materials is "terra sigillata" (earth stamped with a seal), described in the first century AD, which may have been the first patented

© The Author(s), under exclusive license to Springer Nature Switzerland AG 2025   117
K. Randive and P. Godbole, *Medical Geology for Beginners*,
SpringerBriefs in Medical Earth Sciences, https://doi.org/10.1007/978-3-031-82765-5_5

medicine (Abrahams 2005). In this light, Geo-pharmacy emerges as a science that explores how rocks, minerals, and geological processes contribute to therapeutic, medical, and biomedical applications (Randive 2013).

Natural minerals, commonly referred to as "stone drugs," have been a cornerstone of traditional Chinese medicine for over two millennia. Beyond their historical applications, these minerals hold significant promise for addressing contemporary medical challenges such as viral infections, antimicrobial resistance, and the development of new therapeutic and diagnostic tools. The surfaces of metals and minerals, whether natural or synthetically modified, can display unique properties when altered with hydrophilic or hydrophobic and ionic or covalent recognition sites. These surface modifications pave the way for innovative medical strategies, offering properties distinct from those of isolated metal centres (Carter et al. 2021).

The minerals, including hybrid organo-minerals, have reactive cavities and the capacity for dynamic movement of metal ions, anions, and other molecules within their structures. These characteristics enable minerals to interact with viruses, microbes, and macro-biomolecules through multipoint ionic and non-covalent contacts. This capability opens new possibilities for novel therapeutic and biotechnological applications, as minerals can establish multiple interaction points with biological entities, facilitating the development of advanced medical treatments and technologies (Carter et al. 2021).

Essential nutrients, which are naturally occurring elements crucial for efficient metabolism, are vital to the functioning of living organisms, including humans. In humans, there are 14 elements considered essential: calcium, chromium, copper, fluorine, iodine, iron, magnesium, manganese, molybdenum, phosphorus, potassium, selenium, sodium, and zinc. These elements play key roles in various physiological processes, from bone formation to enzyme function, underscoring their importance in maintaining overall health and well-being (Price 2000).

5.2 Therapies and Treatments in Geo-Pharmacy

5.2.1 Balneotherapy

Balneotherapy encompasses a range of methods and practices that utilize scientifically supported mineral-medicinal waters, muds, and natural gases from springs for therapeutic purposes. These treatments are both medically and legally recognized. Balneotherapy serves as a clinically effective complementary treatment for conditions associated with low-grade inflammation and stress. Although the precise biological mechanisms underlying its efficacy remain incompletely understood, research indicates that neuroendocrine and immunological responses, including humoral and cell-mediated immunity, are involved. These responses contribute to the therapy's anti-inflammatory, analgesic, antioxidant, chondroprotective, and anabolic effects, along with neuroendocrine-immune regulation across various conditions. Hormesis, characterized by beneficial effects resulting from exposure to low doses of stressors, likely plays a pivotal role in these effects. Factors such as heat and specific biochemical components like hydrogen sulphide and radon found in mineral-medicinal waters contribute to hormetic responses during balneotherapy. Numerous studies support the notion that the therapeutic benefits of balneotherapy align with the concept of hormesis, affirming its significance in hydrothermal treatments (Gálvez et al. 2018). Balneotherapy provides unique healing experiences at places like the Dead Sea, Kangal hot spring, and Blue Lagoon. Immersing in highly saline water is safe, effective, and enjoyable for recovery, without the need for chemicals or drugs. It offers minimal side effects and low health risks, making it ideal for treating skin conditions like psoriasis and atopic dermatitis (Matz et al. 2003).

5.2.2 Thalassotherapy

"Thalassotherapy," a term derived from the Greek words "thalassa" (sea) and "therapy," is still empirically recommended for patients with certain skin, rheumatic, or respiratory conditions. Broadly defined, thalassotherapy encompasses not only baths in seawater or saltwater but also seaweed or sand baths, sun exposure, inhalation of marine aerosol, and generally any controlled interaction with marine environments and their natural elements for health-promoting purposes (Antonelli and Donelli 2021). Thalassotherapy utilizes seawater, known for its high mineral content, density, and a chemical composition rich in chlorides—primarily sodium, along with magnesium, calcium, potassium, and iodine. Treatments also involve marine peloids (mud or limes), systematic and methodical exposure to sunlight, applications of hot sea sand, and marine climatotherapy, which considers atmospheric factors such as temperature, humidity, wind, and air pressure (Munteanu and Munteanu 2019) (Fig. 5.1).

Fig. 5.1 A thematic diagram showing **a** Balneotherapy and **b** Thalassotherapy

Fig. 5.2 A thematic diagram showing cryotherapy

5.2.3   Cryotherapy

For centuries, humans have utilized cold temperatures for therapeutic, health, and sporting recovery purposes. This application of cold for therapeutic benefits is commonly referred to as cryotherapy. Cryotherapies, which include ice, cold-water, and cold air treatments, are popular due to their ability to remove heat, lower core, and tissue temperatures, and modify blood flow in the body. These effects lead to benefits such as reduced pain perception, known as analgesia, and an improved sense of well-being (Allan et al. 2022). Physiologically, the efficacy of cryotherapy is primarily attributed to its analgesic benefits (Murray and Cardinale 2005) (Fig. 5.2).

5.2.4   Geophagy

Geophagy, or geophagia, is the practice of consuming clay, such as chalk or kaolin. This habit is often associated with materials commonly known as Calabar chalk, ndom, nzu, or Calabar stones (Ijeoma et al. 2014). Physiologically, many nutritionists and researchers consider geophagy to be one of the practices that help provide physical relief from pain or distress. In gastroenterology, clay (kaolin) is believed to absorb toxins from food or bacteria that are associated with stomach upset (Johns and Duquette 1991). Studies also suggest that geophagy may be beneficial by protecting

Fig. 5.3 A thematic sketch showing the practice of consuming clay (Geophagy)

against harmful pathogens and toxins through two distinct physiological pathways (Young and Miller 2019) (Fig. 5.3).

5.2.5 Minerals in Skin Products

In cosmetics production, mineral raw materials like bentonite, kaolin, illite, mica, and talc are used based on their mineral and chemical composition. For instance, clays rich in silicon hydrate tissues and reduce inflammation, while aluminium aids healing, hydration, pigment dispersion, and melanin adsorption. Clays containing silicon, aluminium, calcium, titanium, iron, and potassium offer bactericidal, antiseptic, and regenerative effects, promoting cell renewal and tissue invigoration (Ghaffarian and Muro 2013). Aluminium clays and minerals are commonly found in numerous skin-care products like creams. Additionally, they can be used independently to treat pimples and various skin rashes (Matike et al. 2011). Bentonite massage creams can open the skin pores, aiding the penetration of active minerals like copper (Cu), zinc (Zn), and magnesium (Mg). These minerals play a crucial role in promoting proper regenerative processes in the skin (Murray 2006). Montmorillonite is an active ingredient known for its anti-acne and anti-inflammatory properties. It provides essential mineral elements, stimulates cell activity, eliminates toxins from cells, and aids in the treatment of acne and other skin conditions. Montmorillonite also reduces skin inflammation and tightens the skin. It is commonly used in masks, hair masks, creams, shower gels, soaps, and various other skincare products, often at concentrations of several dozen percent (Elmore 2003; CDC 2010). Fuller's earth, commonly known

Fig. 5.4 A thematic sketch showing application of Fuller's earth (Multani Mitti) for skincare

as Multani Mitti in Urdu, is traditionally utilized in skincare cosmetics, especially for removing blackheads and treating oily skin. It is also employed to enhance skin complexion (Rehan et al. 2019) (Fig. 5.4).

5.3 Ancient Indian Medicinal System: The Ayurveda

Ayurveda, known as the science of life or the science of longevity, aims to promote positive health, prevent diseases, and help individuals achieve long life (Randive 2013). This ancient science, part of the great Vedic tradition, was first recorded around 5000 years ago (Pandey et al. 2013). As one of the oldest scientific medical systems in the world, Ayurveda boasts a long history of clinical experience and a holistic approach to medicine. It not only provides cures for diseases but also offers guidance on maintaining and protecting health. Ayurveda details the diets, medicines, and behaviours that are beneficial or harmful to life. Rasashastra is the discipline within Ayurveda that focuses on the use of minerals and metals for curing diseases. Although initially developed as a separate science, Rasashastra was later integrated into Ayurveda. While it is not considered one of the eight main branches of Ayurveda, it has become an indispensable part of Ayurvedic treatment (Dole and Paranjpe 2004).

The term "Rasashastra" is derived from two Sanskrit words: "Rasa" and "Shastra." "Shastra" translates to science, while "Rasa" has multiple meanings in Sanskrit, including liquid, taste, one of the seven constituents of the body, one of the nine emotions, and cosmos. In the context of this science, however, "Rasa" specifically refers to mercury. Thus, the literal meaning of Rasashastra is the detailed and scientific study of mercury (Kaundal and Arora 2023; Savrikar and Ravishankar 2011).

There appear to be two reasons for designating mercury as "Rasa." First, mercury is the only metal with the unique property of mixing homogeneously with other metals while retaining its liquid state. Second, the potent properties of drugs prepared from mercury, which are believed to prevent death, old age, and pain, may have contributed to this nomenclature (Dole and Paranjpe 2004). Rasashastra comprises two branches: Dhatuvada and Dehavada. Dhatuvada, akin to alchemy, focuses on the techniques for converting non-precious metals into precious metals, such as gold. Dehavada, on the other hand, is concerned with the development of medicines aimed at enhancing longevity, vigor, and vitality. The science of Rasashastra was developed and practiced by adherents of three distinct cults: Buddhists, Shaktas, and Nathpanthis (Savrikar and Ravishankar 2011).

The development of Rasashastra as a science coincided closely with the advancement of mining and metallurgy in ancient India. Four distinct periods can be prominently identified in its evolution (Satpute 2003). Excavations at sites such as Harappa and Mohenjo-Daro during the prehistoric period (4500 BC–1500 BC) have unearthed evidence of sophisticated metal usage. Artifacts including axes, daggers, bow-heads, and utensils crafted from copper, bronze, gold, silver, and lead indicate a well-established metallurgical tradition. Additionally, discoveries of bitumen, cinnabar, gemstones, and other materials suggest a diverse array of craftsmanship. The practical skills of pottery making, painting, and metallurgical techniques such as forging, casting, and smelting were prevalent during this era. (Randive 2013).

During the Vedic period (1500 BC–600 BC), significant advancements were made in metallurgy and chemistry. All four Vedas contain references to various metals and minerals, with the Atharvaveda being particularly notable for its insights into chemical practices for both practical arts and medicinal purposes. Scholars propose that the Atharvaveda may have laid the foundation for later developments in Rasashastra. Similar chemical practices and mineral evidence are also documented in other Vedic texts, including the Upanishads (philosophical writings exploring the nature of reality and self), Srutis (divinely revealed knowledge encompassing the Vedas), and Brahmana Granthas (texts that provide detailed instructions for rituals and sacrifices). These texts collectively illustrate the ancient understanding of the connections between chemistry, minerals, and the broader cosmos (Randive 2013).

The period from 600 BC to 800 AD marks the Ayurvedic era, a pivotal period for Rasashastra. Initially, Ayurveda flourished, resulting in the creation of seminal works such as the Charaka Samhita and Sushruta Samhita. Later, Rasashastra emerged as a distinct discipline within Ayurveda, focusing on the therapeutic application of minerals and metals. Both the Charaka and Sushruta Samhitas contain numerous references to metallic and mineral medicines. Kautilya's Arthashastra (321 BC– 396 BC) significantly contributed to metallurgical knowledge, providing systematic accounts of ore smelting processes and references to fermented fluids and salts (Randive 2013).

The exact origins of advanced Rasashastra are difficult to pinpoint, but scholars agree that its development did not extend beyond the eighth century AD. Texts on Rasashastra began to emerge between the fifth and sixteenth centuries AD, with notable works attributed to this period. Although earlier scriptures may have existed before 800 AD, none have survived. The Rasahridayatantra, attributed to Govinda Bhagavatpada, is considered one of the oldest texts on Rasashastra, dating back to around 800 AD. Numerous texts on Rasashastra were authored between 800 and 1600 AD, many of which are available in published form (Randive 2013).

5.4   Minerals as Medicine

Of all the inorganic substances used in Indian medicine, the most important is mercury. Pure mercury is consumed in different forms for getting long and disease-free healthy life (Zhao et al. 2022). Other substances are classified into three categories as shown in Fig. 5.1.

5.4.1   Maharasa

According to the Rasaratnasamuchchaya, the Maharasas category includes the following substances: Abhraka (mica), Vaikranta (tourmaline), Makshika (chalcopyrite), Vimala (pyrite), Shilajatu (black bitumen), Sasyaka (blue vitriol), Chapala (selenite), and Rasaka (zinc ore) (Rajput et al. 2016). These substances are further elaborated upon in the text.

5.4.1.1   Mica

According to Ayurvedic classification, mica can be categorized based on either its colour or its response to heating. Colour-wise, it can be categorized as white (Sweta), yellow (Pitta), red (rakta), and black (Krishna). Alternatively, based on its response to heating, mica can be classified as pinaka (which sounds like an arrow), naga (which sounds like the hissing of a snake), manduka (which jumps like a frog), and vajra (which is very strong and tough). Among these, the black mica of the vajra type is considered the best (Wele et al. 2021). Incinerated mica is used in various diseases such as diabetes mellitus, dermatoses, tuberculosis, asthma, cough, anaemia, colitis, epilepsy, hysteria, etc. It is believed to promote physical and mental vigour, sexual power, and immunity (Wijenayake et al. 2014).

5.4.1.2 Tourmaline

Ancient texts have described eight varieties of tourmaline based on color: white, red, blue, yellow, gray, blackish, black, and multi-coloured. Among these, black tourmaline is utilized for medicinal purposes. It is employed in the treatment of fever, skin diseases, leprosy, tuberculosis, diabetes mellitus, piles, tumors, and ascites (Hranush and Arakelyan 2020).

5.4.1.3 Chalcopyrite

Ayurvedic texts mention three varieties of chalcopyrite or makshika: Suvarnamakshika (pure chalcopyrite), Raoupyamakshika (white pyrite), and Kansyamakshika (arsenopyrite) (Ranade 2022). Suvarnamakshika bhasma, derived from pure chalcopyrite, is used in the treatment of various diseases such as anemia, anasarca, hyperacidity, burning sensation in the body, diabetes mellitus, piles, dermatosis, bleeding disorders, and more. It is considered a rejuvenating agent and aphrodisiac, beneficial for the eyes, and promotes longevity (Deshpande et al. 2024; Gupta et al. 2010).

5.4.1.4 Pyrite

Similar to chalcopyrite, pyrite is also said to have three varieties: Suvarna vimala, Roupya vimala, and Kansya vimala, among which Kansya vimala is considered the best for medicinal purposes. It is primarily used as a rejuvenator and aphrodisiac (Sharma 2018).

5.4.1.5 Bitumen

Black bitumen, known as Shalajatu, earns its name from its thick exudation from stones heated by the sun in summer, presenting in various shades. It is classified based on odour, such as Gomutra Shilajatu (resembling cow's urine) and Karpura Shilajatu (reminiscent of camphor), and by the metallic ores present in the source area, including Suvarna (gold), Roupya (silver), Tamra (copper), and Loha (iron). Shalajatu primarily acts to purify and strengthen the genitals and urinary tract (Mookerjee 1938). It is utilized as a remedy for numerous ailments like diabetes mellitus, urinary stones, dysuria, anaemia, skin disorders, and edema. Furthermore, it serves as a potent aphrodisiac and rejuvenating agent for patients with tuberculosis, asthma, chronic bronchitis, chronic digestive diseases, nerve disorders, and fractures. Shalajatu is also considered highly effective in treating obesity and cough (Shahriari et al. 2018).

5.4.1.6 Chalcanthite (Blue Vitriol)

Sasyaka, also known as blue vitriol, is described as rejuvenating and detoxifying in Ayurvedic texts. It is particularly beneficial for the eyes and functions as an emetic and purgative agent. Externally, it can be applied as an ointment in conditions like conjunctivitis and trachoma. Additionally, Sasyaka is useful in the treatment of wounds and ulcers with hyper-granulation of tissues. It is also employed for various other conditions including skin diseases, syphilis, gonorrhoea, vitiligo, diabetes, abdominal pains, and worm infestations (Pradhan et al. 2022).

5.4.1.7 Zinc Ores

Zinc ores, including sphalerite, zincite, smithsonite, willemite, and calamine, are collectively referred to as Rasaka in Ayurvedic texts. Although Rasaka represents various zinc ores, each resembling a different type of zinc ore, it is an important drug used in the preparation of a variety of medicines. Rasaka is renowned for its ability to strengthen body tissues and is beneficial in various conditions, including all types of diabetes, diseases of the eyes, asthma, diarrhea, and tuberculosis. Its medicinal properties make it a valuable component in many formulations across Ayurvedic medicine (Panda 2014).

5.4.2 Uparasa

According to Rasashastra, following substances are categorized as Uparasa, (1) Gandhaka (sulfur), (2) Gairika (hematite), (3) Kasisa (melanterite), (4) Kankshi (Alum), (5) Haratala (Orpiment), (6) Manahshila (Realgar), (7) Anjana and (8) Kankushtha (Randive 2013).

5.4.2.1 Sulphur

Ancient Ayurvedic texts classify sulphur into four varieties based on its colour: Shveta (white), Pitta (yellow), Rakta (red), and Krishna (black), with the black variety being considered the best (Singh and Mishra 2019). Sulphur ointment is used for wound dressing due to its antimicrobial properties, making it effective in treating skin diseases such as scabies and various dermatoses (Gupta and Nicol 2004). Additionally, sulphur is used as an oral medicine for treating tuberculosis, loss of appetite, asthma, cough, diseases of the oral cavity, and anorectal diseases (Eick et al. 2009).

5.4.2.2 Hematite

In Ayurveda, hematite is classified into two varieties: Suvarna-gairika and Pashan-gairika. Before use, it is purified with cow's milk or cow's ghee. Hematite is known to mitigate vomiting and hiccups and is beneficial in treating anaemia, menorrhagia, and bleeding diathesis. Externally, it is applied as an ointment to promote wound healing and relieve itching in urticaria when its paste is applied to wounds (Randive 2013).

5.4.2.3 Melanterite

In Ayurveda, melanterite is known in two varieties: the green variety is called Valuka-sisa, and the yellow variety is called Pushpakasisa. Melanterite is used to treat anemia and is effective in managing splenomegaly associated with anemia. It is an important medicine for treating debility, serves as an expectorant, and is also beneficial in treating dysuria and urinary calculi (Randive 2013).

5.4.2.4 Alum

According to Ayurveda, two varieties of alum are recognized: the yellowish one is called Phataki Kankshi, and the snow-white variety is known as Phullika Kankshi. Purified alum is utilized to stop bleeding. Alum paste is applied in tinea versicolor, a skin disease, while eye drops made from alum are beneficial in treating eye diseases. Alum diluted in water is used as a vaginal douche. Additionally, alum water is administered orally in cases of poisoning. Its astringent property helps delay the absorption of toxins from the gastric mucosa (Randive 2013).

5.4.2.5 Orpiment

Ayurvedic texts describe three varieties of orpiment: Patra Haratala (layered form), Pinda Haratala (fragmentary form), and Tabaki Harata (prepared artificially, extremely toxic, and not used for medicinal purposes). Haratala Bhasma, derived from orpiment, is considered the choicest remedy for various skin diseases. It is also beneficial in treating gout, syphilis, and gonorrhea. Furthermore, it is recognized as an aphrodisiac, improves complexion, and contributes to longevity (Shubh and Hiremath 2010).

5.4.2.6   Realgar

Ayurvedic texts describe three varieties of realgar: *Shyamangi*, characterized by a blackish-red hue with a slight yellowish tint; *Kanaviraka*, exhibiting a copper-like reddish colour and a shiny appearance without any yellow spots; and *Khandakhya*, distinguished by its bright color, heaviness, and ease of powdering. Realgar is employed in the treatment of conditions such as bronchial asthma, tuberculosis, cough, fever, and itching (Randive 2013).

5.4.2.7   Stibnite, Galena, and Other Minerals

Ayurvedic texts detail five varieties of Anjana, including Srotonjana and Sauviranjana, both comprising stibnite, Rasanjana made of yellow oxide of mercury, Pushpanjana, a blend of specific flowers, alum, and zinc, and Nilanjana, derived from galena. Anjana, commonly known as Surma in Hindi, is a soft black substance used in eye care. Rasanjana and Pushpanjana are predominantly employed in treating ocular ailments, while Nilanjana is favoured for cosmetic applications (Randive 2013).

5.4.3   Sadharana Rasa

In Rasashastra, the Sadharana Rasa category encompasses several substances, including Kampilla, Gauripashaha (arsenolite), Navasagar (sal-ammoniac, ammonium chloride), Kapardika (cowrie), Vahnijara, Girisindura (montroydite), Hingula (cinnabar), and Mriddarshringa (litharge, lead monoxide). Notably, Kapilla, Vhanijara, and Kapardika are exceptions to this grouping as they are not minerals (Randive 2013).

5.4.3.1   Arsenolite

The Ayurvedic varieties of Gaouripashana or Somala are Sphatikabha, Shankhabha, and Pitabha. The purified mineral serves as an antidote for scorpion bites and is also effective in treating joint disorders and syphilis (Gandhi and Ingole 2023).

5.4.3.2   Ammonium Chloride or Sal-Ammoniac

Two varieties of Navasagar are recognized: Yogambari and Chullika. Classified as a salt, Navasagar is employed for the purification of gold. Internally, it serves as a medicine for conditions like guinea worm infection, also known as Ion Dracunculiasis. Additionally, Navasagar is used as an expectorant to relieve mucous (Akshata and Raghuveer 2015).

5.4.3.3 Montroydite

The mineral described is an alteration product of cinnabar. It acts as a potent purgative and is known for its ability to strengthen the body. It is particularly beneficial for the eyes and is used topically for wound dressing. Additionally, it is employed in the treatment of skin conditions such as eczema and pruritus, often in the form of an ointment (Hocsman et al. 2006; Randive 2013).

5.4.3.4 Cinnabar

According to Ayurvedic texts, there are two varieties of Hingula: Shukatunda, which is bright red, and Hanspada, characterized by red color with white streaks. Hingula is considered rejuvenative and is known to enhance appetite. It is also believed to alleviate various ailments such as diabetes, skin diseases, fever, hepatitis, and splenomegaly (Randive 2013).

5.4.3.5 Litharge

According to Ayurveda, there are three varieties of Mriddarshringa: Pita (yellow), Pitapandura (yellowish-white), and Kritrim (artificial). Its ointment is beneficial in treating skin diseases, particularly those associated with itching. Additionally, its powder is useful in healing syphilitic ulcers. Moreover, Mriddarshringa is beneficial as a hair dye.

5.5 Metals as Medicine

While the roots of Rasa Shastra trace back to ancient texts of Indian civilization, its evolution as an independent therapeutic system began around the eighth century A.D. Ayurvedic classics such as the *Charaka Samhita* and *Sushruta Samhita*, predating this period, include detailed descriptions of metals, their processing methods, and their applications in therapeutics (Galib et al. 2011). Formulations containing mercury are seldom mentioned in Charaka Samhita. The first reference regarding Parada and its therapeutic utility in this classic is subject to controversy. Only a few scholars interpret the term "Rasa" in the verse chikitsasthana 7/71 as referring to Parada. The second reference is found in Dwivraniya Chikitsa, where the term "Rasa" is interpreted as Parada by the commentator Chakrapani (Caraka 2000; Cakrapani on Caraka 2000). *Swarna*, also referred to as *Sara Lauha*, stands as a significant and esteemed metal recognized by Indians since ancient times. A multitude of formulations containing 'Swarna' prove beneficial, serving various purposes such as Vrishya (aphrodisiac), Balya (tonic), Rasayana (rejuvenative), Medhya (nervine tonic), Ayushya (longevity promoter), Ojo Vardhaka (immunomodulator), and Vayah

Sthapaka (anti-aging). These formulations also act as disease-alleviators, particularly in chronic debilitating conditions like Raja Yakshma (tuberculosis), Swasa (asthma), Kasa (cough), and Pandu (anaemia). The recommended dose for 'Swarna Bhasma' typically ranges from 15 to 30 mg (Rasa Vagbhatta 1998; Sadnanda Sharma15/69–71, 15/81 1998; Somadeva 2004). Rajata (Silver), akin to gold as a noble metal, has captivated the interest of ancient Acharyas. The utilization of silver in therapeutics traces back to the era of Charaka and his contemporaries. While its therapeutic applications may not be as extensive as other metals like Tamra or Loha, ancient classics unveil that silver held significance in Ayurvedic therapeutics (Sadanand Sharma 15/18 1998). Tamra (Copper) stands as another ancient metal renowned in human civilization. Formulations containing 'Tamra' (Copper) are effective in treating various diseases such as worm infestation, obesity, haemorrhoids, consumption, anaemia, skin diseases, asthma, cough, hyperacidity, inflammation, pain, liver disorders, and gastrointestinal issues. Additionally, Charaka advocates the use of Tamra Patra (copper vessels) in pharmaceutical procedures (Sadanand Sharma 17/46 1998). After Swarna (Gold), Rajata (Silver), and Tamra (Copper), Loha (Iron) or Ayasa is another metal known to ancient civilizations. Various formulations of 'Loha' are useful in a wide range of diseases: Sula (Pain), Arsha (Hemorrhoids), Gulma (Abdominal Tumors), Pliha Roga (Spleen Disorders), Yakrit Roga (Liver Disorders), Ksaya (Consumption), Pandu (Anaemia), Kamala (Jaundice), etc. (Rasa Vagbhatta 1998). Mandura, the second form of Iron, has been utilized in a wide array of therapeutic procedures in classical Ayurveda since ancient times (Madhava Upadhyaya 1999). Naga is a significant form of Puti Loha known since ancient times, also identified by other terms like Sisaka or Sisa. Charak emphasizes that the medicinal uses of this metal should be external, particularly in cases of Mandala Kusta (Somadeva 2004). Vanga (Tin), one of the Puti Lohas, was recognized by ancient Indian physicians under the name Trapu. In Charaka Samhita, this metal is categorized under Parthiva Dravyas. Formulations containing 'Vanga' are beneficial in treating various diseases such as Prameha (urinary disorders), Kasa (cough), Shwasa (asthma), Krimi (worm infestation), Ksaya (consumption), Pandu (anemia), Pradara (menorrhagia), and Garbhashaya Chyuti (miscarriage) (Sadanand Sharma 18/39–42 1998). Kamsya, another significant Misra Loha, is an alloy of Copper and Tin known since the period of Samhita Kala. Formulations containing 'Kamsya' are beneficial in treating diseases like Krimi (worm infestation) and Kusta (skin diseases) (Rasa Vagbhatta 1998). Medicinal uses of some essential metals is given in Table 5.1

Table 5.1 Medicinal uses of some essential metals (after Randive 2013)

Name	Taste	Post digestive effect	Potency	Attributes	Doshas alleviated	Therapeutic uses	Dose
Gold (Suvarna)	Sweet astringent bitter	Sweet	Cold	Unctuous rejuvenative	Vata, Pitta	Rejuvenating, augments memory, semen, intelligence, strength. Detoxifier, strengthens heart, vessels	30–120 mg
Silver (Roupya)	Sour astringent	Sweet	Cold	Unctuous, nervine	Vara, Pitra	Nervine, tonic works well in paralysis, diseases of eyes and muscles	30–60 mg
Copper (Tamra)	Bitter astringent	Sweet	Hot	Sharp	Kapha	Enlargement of liver and spleen, anaemia colitis, anasarca, ascites	60–120 mg
Iron (Loha)	Astringent	Sweet	Cold	Sharp	Kapha, Pitta	Rejuvenator, anaemia, piles, dermatoses, liver and spleen diseases	125–250 mg
Tin (Vanga)	Astringent	Pungent	Hot	Hot, dry sharp	Kapha, Vata	Diabetes, sexual debility dysmenorrhoea, dermatoses	60–240 mg
Lead (Naga)	Bitter	Pungent	Hot	Unctuous	Vara, Kalpa	Urinary and skin diseases, diabetes, tumours	60–120 mg

(continued)

Table 5.1 (continued)

Name	Taste	Post digestive effect	Potency	Attributes	Doshas alleviated	Therapeutic uses	Dose
Zinc (Jasada)	Astringent bitter	Sweet	Cold	Dry Pitta	Kapha, Tonsillitis, Colitis	Diseases of eyes, diabetes	60–240 mg
Trivanga (No 5 + 6 + 7) combination	–	–	–	–	–	Secondary sterility in males and females, abortion, sexual debility, diabetes	120–240 mg

5.6 Medicinal Value of Gemstones and Semi-Precious Stones

Ancient Ayurvedic texts of Rasashastra mention the use of gemstones for medicinal purposes. The study of gemstones initially aimed to stabilize mercury, alleviate diseases, enhance longevity, avert ill effects of planets, and use them in medicinal preparations. Gemstones are classified in Ayurveda based on their preciousness: Ratna (precious), Upratna (semi-precious), and Kshudra Ratna (non-precious stones). The five superior gemstones are Manikya or Padmaraga (Ruby), Indranila or Nilam (Blue Sapphire), Tarkshya (Emerald), Pushparaga (Yellow Sapphire), and Vajra or Hirak (Diamond). Other gemstones include Mukta (Pearl), Pravala (Coral), Gomeda (Hessonite), and Vaidurya (Cat's eye), collectively known as Nav-Ratnas (nine gemstones). Semi-precious stones include Vaikrant (tourmaline), Suryakanta (spinel), Chandrakanta (moonstone), Rajavarta (Lapis Lazuli), Perojaka (turquoise), Sphatika (rock crystal), Vyomashma (jade), Palanka (onyx), Rudhira (carnelian), Puttika (peridot), and Trinakanta (amber). These stones are used as medicines in the form of fine powder (Pishti) or incineration (Bhasma) (Randive 2013).

5.7 Therapeutic Value of Crystals

Crystals, amulets, and gemstones have been utilized across various cultural and healing traditions worldwide for healing, protection, decoration, and adornment. While evidence of their earliest usage is challenging to pinpoint, some texts suggest that the origins of crystals can be traced back to healing rites in ancient Egypt and Mesopotamia. Stones like lapis lazuli and malachite were notably employed for their healing and protective properties in these early civilizations (McClean 2013). The specific placement of crystals can influence our well-being due to their ability to hold and emit energy vibrations. Just as crystals are utilized in watches to regulate time, they can interact with our electromagnetic energy fields or subtle bodies, which encompass and permeate the physical body. These subtle bodies include the etheric, emotional, and mental bodies, collectively known as the aura. Crystals absorb, focus, direct, and diffuse energy fields, assisting a diseased or imbalanced body in finding its natural energetic rhythm. Placing appropriate crystals on the seven main chakras, which are spinning wheels of subtle energy aligned along the centre of the torso, can connect the aura's energy fields with emotions, glands, organs, physical body parts, and circulatory flows. For instance, wearing yellow citrine may uplift one's mood, rose quartz may help ease heartache, and amethyst may calm a busy mind and aid in sleep. This suggests that crystals have a tangible impact on our well-being by interacting with our subtle energy systems (Dubey 2019).

Vedic texts, such as the Garuda Purana and Graha-gocara Jyautisha, provide valuable insights into the significance of crystals in India. According to these texts, the origin story of ayurvedic crystal healing practices is intertwined with the Vedic demon Vala. Legend has it that Vala's body parts became gem seeds after being dismembered by Hindu Demigods during a sacrificial ritual. These gem seeds fell into various natural elements, such as rivers, oceans, forests, and mountains, imbuing them with talismanic powers. The Ruby, believed to be the spilled blood of demon Vala, holds talismanic powers related to blood circulation and courage. Pearls, representing Vala's teeth, are used for calming the mental state. Yellow Sapphire, symbolizing Vala's skin, promotes general well-being and aids in pregnancy stages. Hessonite, derived from Vala's fingernails, is believed to avert disaster and insanity and aid those in scientific fields. Emerald, associated with Vala's bile, enhances psychic powers, learning ability, and clairvoyance. Diamond, symbolizing Vala's bone, induces purity, creativity, and happiness while aiding in diseases of the sex organs. Cat's Eye, representing Vala's war cry, assists in protection from enemies, drowning, intoxication, and government punishment. Blue Sapphire, signifying Vala's eyes, wards off the evil eye and protects travellers. Coral, symbolizing Vala's intestines, helps remove obstacles and resolves financial issues. Red Garnet, representing Vala's toenails, shares similar properties with Ruby. Jade, derived from Vala's scattered fat, aids in karma removal and shares properties with Emerald. Rock Crystal, symbolizing Vala's semen, shares powers with Pearl, and Bloodstone, reflecting Vala's complexion, shares properties with Coral. The significance of each crystal is enhanced by its source on Vala's body, and Ayurvedic practices assign these stones to corresponding chakras, invoking their healing properties for holistic well-being (Carlos 2018) (Tables 5.2 and 5.3).

Table 5.2 Properties of gemstones (after Randive 2013)

No	Name and chemical composition	Structure	Hardness	S.G.*	R.I*	D.R.*
1	Ruby (Manikya) Al,O	Trigonal	9	4.00	J.76–1.77	0.008
2	Pearl (Mouktika) CaCO, $C_3H_{18}N_9O_{11}nH_2O$	Orthorhombic	3	2.71	153–1.68	N/A
3	Coral (Pravata) CaCO, or $(C_3H_{48}N_9O_{11})$	Trigonal	3	2.68	1.49–1.66	N/A
4	Emerald (Tarkshya) $Be_3Al_2 (SiO_3)_6$	Hexagonal	7.5	2.71	1.57–1.58	o.oo6
5	Yellow Sapphire (Pushparaga) Al_2O_3	Trigonal	9	4.00	1.76–1.77	0.008
6	Diamond (Vajra, Hiraka) C	Cubic	10	3.52	2.42	None
7	Blue Sapphire (Nila) Al_2O_3	Trigonal	9	4.00	1.76–1.77	0.008
8	Hessonite (Gomeda) $Ca_3Al(SiO_4)_3$	Cubic	7.5	3.65	1.73–1.75	None
9	Cat's-Eye (Vaidurya) $BeAl_2O_4$	Orthorhombic	8.5	3.71	1.74–1.75	0.009

*S.G Specific gravity. *RI Refractive Index. *D.R Double refraction or Birefringence

Table 5.3 Properties of semi-precious gemstones (after Randive 2013)

No	Name and chemical composition	Structure	Hardness	S.G	R.I	D.R
1	Fluorite (Vaikrant) CaF_2	Cubic	4	3.18	1.43	None
2	Spinel (Suryakant)	Cubic	8	3.60	1.71–1.73	None
3	Moonstone (Chandrakant) $KAISi_3O_8$	Monoclinic	6	2.57	1.52–1.53	0.005
4	Lapis Lazuli (Rajavarta) (Na, $Ca)_8(Al,Si_{12}O_{24}(SO_4)Cl_2(OH)_2$	Various	5.5	2.80	1.50 (mean)	None
5	Turquoise (Peroja) $CuAl_6(PO_4)$, $(OH)_8$ $5H_2O$	Triclinic	6	2.80	1.61–1.65	0.004
6	Rock Crystal (Sphatika) SiO_2	Trigonal	7	2.65	1.54–1.55	0.009
7	Jade (Vyomashma)Na (Al, Fe) Si_2O6	Monoclinic	7	3.33	1.66–1.68	0.012
8	Onyx (Palanka) SiO2	Trigonal	7	2.61	1.53–1.54	0.004
9	Carnelian (Rudhira) SiO2	Trigonal	7	2.61	1.53–1.54	0.004
10	Peridor (Puttika) $(Mg,Fe)_2SiO_4$	Orthorhombic	6.5	3.34	1.64–1.69	0.036
11	Amber (Trunakant) Mainly $C_{10}H_{16}O$	Amorphous	2.5	1.08	1.54–1.55	N/ A

5.8   Calcium from Geological Sources

Calcium compounds, classified as Sudha varga in Rasashastra, include various
mineral and animal sources such as quick lime, chalk, talc, selenite, and others
like conch and coral. Quick lime, made from limestone, aids digestion, alleviates
abdominal pain, and treats colitis. Chalk pacifies bile, treats burning sensations,
blood disorders, and diarrhoea. Talc powder improves taste sensation, relieves fever,
abdominal pain, and skin burning, while also treating bleeding piles and dysentery.
Selenite powder is cooling, treats bile disorders, bleeding, hyperacidity, and calcium
deficiency disorders like rickets and osteoporosis. It also aids growth in children,
strengthens teeth, and controls bleeding in menorrhagia and leucorrhoea (Randive
2013).

5.9   Medicinal Uses of Fossils

Fossil Encrinite or Fossil Norinite, also known as Badarashma or Ashmabhid, is
utilized in Unani Medicine as Hajaratbera. Resembling jujube fruit in shape, it effectively treats urinary disorders like burning sensation, frequency, and difficulty of
urination. It is particularly beneficial for conditions like oliguria, anuria, dysuria,
and urinary stones, possessing diuretic and lithotopic properties. Oyster shells,
known as Shukti, include varieties like muktashukti and Jalashukti. Jalashukti stimulates appetite, aids digestion, and reduces intoxication, while muktashukti improves

appetite, reduces urine sugar, and is used in heart diseases, asthma, and colics. Incinerated gastropod shells, or Shankha bhasma, are hot and astringent, ideal for hyperacidity, peptic ulcers, digestion improvement, and acne treatment. Gastropod shells from the Cypraiedae family, known as cowrie, have similar uses but are processed differently due to their small size and unique shape. Coral, incinerated as Chandraputi Pravalbhasma, is beneficial for bile disorders, while Suryaputi and agniputi Pravalbhasma are used for cough-related ailments. Generally, coral relieves burning sensations, high fever, and bleeding from various body parts (Randive 2013).

5.10   Mercurial Medicines of Rasashastra

The ancient and modern texts of Rasashastra mention the geological materials discussed earlier, primarily in forms such as incineration, pastes, pulps, tablets, and syrups. Mercury serves as the base for most of these formulations, as highlighted earlier. Here are some of the most important mercurial formulations used in Rasashastra for the awareness of medical geologists.

5.10.1   Kajjali

Kajjali is the fine jet-black powder obtained by triturating mercury with sulphur and minerals, without adding any liquids. It serves as a common adjunct or basic ingredient in various formulations and can also be used as a medicine on its own (Joshi et al. 2021).

5.10.2   Parpati

The Sanskrit term "Parpata" refers to a thin, crisp wafer, while "Parpati" describes a preparation that is thin, brittle, and shaped like a thin crisp wafer. Parpati's main ingredient is Kajjali, which is mixed with other ingredients, rubbed, and then transferred into an iron vessel coated with ghee. It is gently heated until the drug substance melts, poured onto a banana leaf coated with ghee, and pressed between another ghee-coated banana leaf. After cooling, the upper leaf is removed, and the Parpati drug is separated, washed with hot water, dried, powdered, and bottled. Parpati preparations are commonly used to treat diseases caused by ama in the gut, such as anorexia, colitis, gout, piles, diarrhoea, dysentery, and bronchial asthma (Sud and Bandari 2014).

5.10.3 Rasapushpa

Equal quantities of purified mercury, pure ferrous sulphate, and rock salt are meticulously rubbed together until a homogeneous mixture is achieved. This blend is then placed in a glass bottle, which is covered with multiple layers of mud-cloth. After thorough drying, the mixture undergoes gradual heating. The white flakes that form at the neck of the bottle upon cooling are collected, constituting Rasapushpa. Rasapushpa serves as a purgative and is employed in treating conditions like hiccups, syphilis, among others.

5.10.4 Raskarpura

Purified mercury and rock salt are combined and triturated before being processed with the juice of Snuhi (Euphorbia neriifolia). The mixture is extensively rubbed, dried, and stored in a sealed iron vessel, which is then heated gradually for 12 h in a Lavan yantra. After cooling, the sealing is removed, and the white calx is collected from the upper part of the vessel. Rasakarpura, although highly toxic, exhibits properties such as anti-diarrheal, anti-dermatosis, and antimicrobial effects, serving as an alternative blood purifier. It finds utility in treating dysentery, skin diseases, and syphilis, but caution must be exercised due to its toxicity, and it should be used in very small doses (Sud 2016).

5.10.5 Rasasindura (Red Mercury Sulfide)

Two parts of purified mercury, one-fourth part of purified sulfur, and one-sixteenth part of purified navasagar (ammonium chloride) are meticulously rubbed together in a stony mortar and further processed with the juice of jambira (Citrus limonum). The resulting powder is then filled into a bottle, which is covered with rags and mud. After drying, the bottle is placed in a Valuka yantra where sand is filled up to the neck of the bottle and subjected to gradual heating for 24 h. Upon cooling, the bottle is broken, and a bluish Rasasindura is collected at the neck. Rasasindura is highly esteemed as a treatment for respiratory ailments such as bronchitis, pneumonia, pleurisy, dyspnoea, chronic cold, and asthma (Biswas and Bellare 2022).

5.10.6 Makaradhvaja and Siddhamakaradhvaja

One part of fine gold sheets and eight parts of purified mercury are meticulously rubbed together until the gold integrates completely with the mercury. Then, sixteen parts of purified sulfur are added and further rubbed together to form Kajjali. This mixture is processed with the juice of red cotton plant (Gossypium herbaceum) and Aloe Vera. The resulting mixture is filled into a bottle and subjected to the same procedure mentioned earlier for Rasasindura. Siddhamakardhvaja is prepared by adding camphor to the mixture. Both Makardhvaja and Siddhamakardhvaja are esteemed as potent aphrodisiac preparations and are utilized for tissue depletion. They are renowned for their rejuvenating properties, delaying the aging process, and preventing issues like wrinkling and hair loss (Khedekar et al. 2011).

5.10.7 Hingulottha Rasa

Cinnabar powder is mixed with the juice of jambira (Citrus limonium) and rubbed for 24 h to form a paste. This paste is then applied to the inner surface of the lower pot of the Urdhvapatan yantra and subjected to heat. As a result, a thin layer of mercury forms on the inner surface of the upper pot. This mercury, obtained through this process, is regarded as the purest form suitable for medicinal use (Shukla and Singh 2019).

5.10.8 Mrittyunjaya Rasa

A mixture containing purified cinnabar, mercury, sulfur, Vatsnabh (Aconitum ferox), Maricha (Piper nigrum), and borax in equal parts is prepared by rubbing them together in ginger juice. This results in a fine semi-solid pulp, from which small tablets weighing 30 mg each are made. When taken with honey, this preparation is effective in alleviating all types of fevers (Jagtap 2019).

5.10.9 Rajamriganka Rasa

A remedy for tuberculosis is prepared using purified mercury (Rasasindura) in 3 parts, along with calx of gold and copper in equal parts, and purified orpiment, realgar, and sulphur in 2 parts each (Debnath 2012).

5.10.10 Hemagarbha Pottali

Kajjali, made from purified mercury, sulfur, gold calx, and copper calx, is mixed with Aloe Vera juice and formed into small pyramid-shaped matras. These are then wrapped in a silk cloth layered with sulfur powder, sealed into a pouch, and heated gradually. After cooling, the matras are used to treat conditions like cough, dyspnoea, tuberculosis, and colitis (Singh and Kailoria 2022).

5.10.11 Vatakuntala Rasa

This medication, composed of musk, purified realgar, purified mercury, sulfur, and various herbs, is employed in the treatment of epilepsy, paralysis, and other ailments (Randive 2013).

5.10.12 Kumarkalyana Rasa

This preparation, combining Rasasindura, Muktapishti, Suvarnamakshika Bhasma, incinerated gold, mica, and iron, processed with Aloe Vera juice, is known as Kumarakalyana. It's particularly helpful for infants and children, addressing issues such as loss of appetite, vomiting, fever, dyspnoea, hepatitis, debility, and diarrhoea (Randive 2013).

5.10.13 Garbhapal Rasa

This drug, prepared from purified Cinnabar, incinerated ashes of lead, tin, and iron, along with various herbs, is aptly named Garbhapal Ras. It is specifically beneficial during pregnancy, starting from the first month of gestation. Garbhapal Ras helps alleviate and prevent diseases during pregnancy while promoting proper foetal growth (Mishra et al. 2012).

5.10.14 Laxmivilas Rasa

Equal quantities of various minerals, including purified mercury, orpiment, realgar, zinc carbonate, and incinerated ashes of lead, copper, mica, bronze, and iron, are mixed and made homogeneous. This mixture is then combined with herbs and herbal juices to form a paste, which is dried and formed into tablets. This preparation is effective in alleviating a wide range of diseases such as tuberculosis, cough, asthma, fever, jaundice, anaemia, edema, abdominal colic, piles, and diabetes mellitus (Gokarn et al. 2015).

5.10.15   Chandrakant Rasa

Equal parts of incinerated mercury, mica calx, iron calx, copper calx, and purified sulphur are combined and rubbed with the latex of Snuhi (Euphorbia nerifolia) for a day. The mixture is then formed into tablets and stored in an iron container. This formulation is beneficial for alleviating sinusitis and other head disorders (Randive 2013).

References

Abrahams PW (2005) The involuntary and deliberate (geophagy) ingestion of soil by humans and other members of the animal kingdom. In: Selinus et al (eds) Essentials of medical geology. Elsevier, N.Y, pp 435–458

Akshata R (2015) A comprehensive review on Navasadara. AAMJ 1(1)

Allan R, Malone J, Alexander J, Vorajee S, Ihsan M, Gregson W, Kwiecien S, Mawhinney C (2022) Cold for centuries: a brief history of cryotherapies to improve health, injury and post-exercise recovery. Eur J Appl Physiol 122(5):1153–1162. https://doi.org/10.1007/s00421-022-04915-5. Epub 2022 Feb 23. PMID: 35195747; PMCID: PMC9012715

Antonelli M, Donelli D (2021) Thalassotherapy, health benefits of sea water, climate and marine environment: a narrative review. In: Proceedings of the 6th international electronic conference on water sciences, 15–30 November 2021, MDPI: Basel, Switzerland. https://doi.org/10.3390/ECWS-6-11606

Biswas S, Bellare J (2022) Explaining Ayurvedic preparation of Rasasindura, its toxicological effects on NIH3T3 cell line and zebrafish larvae. J Ayurveda Integr Med 13(2):100518. https://doi.org/10.1016/j.jaim.2021.08.011. Epub 2021 Nov 29. PMID: 34857444; PMCID: PMC8728081

Caraka (2000) 'Caraka Samhitaa'. Chikitsa Sthaana 7/71. Varanasi, India: Choukhambha Sanskrit Sansthaan

Carlos KD (2018) Crystal healing practices in the western world and beyond. Honors undergraduate theses, 283. https://stars.library.ucf.edu/honorstheses/283

Carter OWL, Xu Y, Sadler PJ (2021) Minerals in biology and medicine. RSC Adv 11(4):1939–1951. https://doi.org/10.1039/d0ra09992a. PMID: 35424161; PMCID: PMC8693805

Centers for Disease Control and Prevention (CDC) (2010) Mercury exposure among household users and nonusers of skin-lightening creams produced in Mexico-California and Virginia. MMWR 61:33–36

Debnath PK, Chattopadhyay J, Mitra A, Adhikari A, Alam MS, Bandopadhyay SK, Hazra J (2012Jul) Adjunct therapy of Ayurvedic medicine with anti tubercular drugs on the therapeutic management of pulmonary tuberculosis. J Ayurveda Integr Med 3(3):141–149. https://doi.org/10.4103/0975-9476.100180.PMID:23125511;PMCID:PMC3487240

Deshpande K, Dhok S, Godbole P, Jangale K, Jawadand S, Randive K (2024) Medicinal applications of Swarna-Makshika, Raupya-Makshika, and Vimala. Int J Mul Res Rev 3(4):57–79, https://doi.org/10.56815/IJMRR.V3I4.2024/57-79

Dole VA, Paranjape P (2004) A Textbook of Rasashastra. Chaukhamba Sanskrit Pratishthan, New Delhi, ISBN: 8170842290 PB; 438

Dubey SR (2019) Crystal therapy. Asian J Nurs Educ Res 9(3):460–462. https://doi.org/10.5958/2349-2996.2019.00095.8

Eick F, Maleta K, Govasmark E, Duttaroy AK, Bjune AG (2009) Food intake of selenium and sulphur amino acids in tuberculosis patients and healthy adults in Malawi. Int J Tuberc Lung Dis 13(10):1313–5. Erratum in: Int J Tuberc Lung Dis 13(12):1579 (2009). PMID: 19793440

Elmore AR (2003) Final report on the safety assessment of aluminum silicate, calcium silicate, magnesium aluminum silicate, magnesium silicate, magnesium trisilicate, sodium magnesium silicate, zirconium silicate, attapulgite, bentonite, Fuller's earth, hectorite, kaolin, lithium magnesium silicate, lithium magnesium sodium silicate, montmorillonite, pyrophyllite, and zeolite. Int J Toxicol 22:37–102

Finkelman RB (2006) Health benefits of geologic materials and geologic processes. Int J Environ Res Pub Health 3(4):338–342. https://doi.org/10.3390/ijerph2006030042

Galib BM, Mashru M, Jagtap C, Patgiri BJ, Prajapati PK (2011) Therapeutic potentials of metals in ancient India: a review through Charaka Samhita. J Ayurveda Integr Med 2(2):55–63. https://doi.org/10.4103/0975-9476.82523. PMID: 21760689; PMCID: PMC3131772

Gálvez I, Torres-Piles S, Ortega-Rincón E (2018) Balneotherapy, immune system, and stress response: a hormetic strategy? Int J Mol Sci 19(6):1685. https://doi.org/10.3390/ijms19061685. PMID: 29882782; PMCID: PMC6032246

Gandhi P, Ingole RK (2023) Arsenicals review: poison vis-a-vis medicine. Int J Ayurveda Pharma Res 11(1):73–81. https://doi.org/10.47070/ijapr.v11i1.2625

Ghaffarian R, Muro S (2013) Models and methods to evaluate transport of drug delivery systems across cellular barriers. JoVE 80:50638. [Google Scholar] [CrossRef]

Gokarn R, Rathi B, Rajput D (2015) Pharmaco-therapeutic profile of an Ayurvedic herbo-mineral formulation Laxmivilas Rasa. J Ind Syst Med 3:141–148

Gupta AK, Nicol K (2004) The use of sulfur in dermatology. J Drugs Dermatol 3(4):427–431. PMID: 15303785

Gupta RK, Lakshmi V, Mahapatra S, Jha CB (2010) Therapeutic uses of Swarnamakshika Bhasma (A Critical Review). AYU 31(1):106–110. https://doi.org/10.4103/0974-8520.68191. PMID: 22131694; PMCID: PMC3215311

Hranush H, Arakelyan H (2020) Black tourmaline and health

Hocsman A, Di Nezio S, Charlet L, Avena M (2006) On the mechanisms of dissolution of montroydite [HgO(s)]: dependence of the dissolution rate on pH, temperature, and stirring rate. J Colloid Interface Sci 297(2):696–704. https://doi.org/10.1016/j.jcis.2005.11.020. Epub 2005 Dec 19. PMID: 16360665

Ijeoma KH, Onyoche OE, Uju OV, Chukwuene IF (2014) Assessment of heavy metals in edible clays sold in Onitsha Metropolis of Anambra State, Nigeria. Br J Appl Sci Tech 4:2114–24. https://doi.org/10.9734/BJAST/2014/7946

Jagtap G (2019) Review of literature of Mrityunjaya Rasa A Herbo-mineral Ayurvedic formulation. Natl J Res Ayurved Sci 5 https://doi.org/10.52482/ayurlog.v7i03.344

Johns T, Duquette M (1991) Detoxification and mineral supplementation as functions of geophagy. Am J Clin Nutr 53:448–456. https://doi.org/10.1093/ajcn/53.2.448

Joshi N, Dash MK, Upadhyay C, Jindal V, Panda PK, Shukla M (2021) Physico-chemical characterization of kajjali, black sulphide of mercury, with respect to the role of sulfur in its formation and structure. J Ayurveda Integr Med 12(4):590–600. https://doi.org/10.1016/j.jaim.2021.05.006. Epub 2021 Nov 10. PMID: 34772584; PMCID: PMC8642700

Kaundal P, Arora A (2023) A review on the principles of Rasa Shastra in Indian system of medicine and its homology with modern chemical processes. J Ayurveda Int Med Sci 12(8):198–203. Available From https://jaims.in/jaims/article/view/2836

Khedekar S, Patgiri BJ, Ravishankar B, Prajapati PK (2011) Standard manufacturing process of Makaradhwaja prepared by Swarna Patra–Varkha and Bhasma. Ayu 32(1):109–115. https://doi.org/10.4103/0974-8520.85741.PMID:22131768;PMCID:PMC3215406

Limpitlaw UG (2004) The medical uses of minerals, rocks, and fossils. In: Geological society of America 2004 annual meeting, abstracts with programs, abstract, vol 36, no 5, pp 48-5

Matike DME, Ekosse GIE, Ngole VM (2011) Physico-chemical properties of clay soils used traditionally for cosmetics in Eastern Cape. S Afr Int J Phys Sci 6:7557–7566

Matz H, Orion E, Wolf R (2003) Balneotherapy in dermatology. Dermatol Ther 16(2):132–140. https://doi.org/10.1046/j.1529-8019.2003.01622.x. PMID: 12919115

McClean S (2013) The role of performance in enhancing the effectiveness of crystal and spiritual healing. Med Anthropol 32(1):61–74. https://doi.org/10.1080/01459740.2012.692741

Mishra D, Sinha M, Dwivedi M, Mishra P, Kumar V (2012). Clinical study of 'Garbhpal Ras' for metals determination, biochemical and hematological profiling in feto-maternal serum. Sri Lanka J Indig Med 2:120–126

Mookerjee B (1938) Rasa Jala Nidhi, vol 2: Minerals (uparasa), ISBN-10: 8170305829, ISBN-13: 9788170305828

Munteanu C, Munteanu D (2019) Thalassotherapy today. Balneo Res J 10:440–444. https://doi.org/10.12680/balneo.2019.278

Murray HH (2006) Bentonite applications. Dev Clay Sci 2:111–130

Murray A, Cardinale M (2015) Cold applications for recovery in adolescent athletes: a systematic review and meta-analysis. Extrem Physiol Med. 4:15

Pandey MM, Rastogi S, Rawat AK (2013) Indian traditional ayurvedic system of medicine and nutritional supplementation. Evid Based Complement Alternat Med 2013:376325. https://doi.org/10.1155/2013/376325. Epub 2013 Jun 23. PMID: 23864888; PMCID: PMC3705899

Pradhan A, Gopikrishna M, Sashidhar (2022) Conceptual review on Sasyaka (Tuttha). https://doi.org/10.20959/wjpr20223-23398

Price WA (2000) Nutrition and physical degradation, 6th edn. The Price-Pottenger Nutrition Foundation, Inc: La Mesa, CA, p 524

Ranade M (2022) Preparation, physio-chemical characterization, and pharmaceutical standardization of a copper based Indian traditional drug Swarna makshik Bhasma prepared by two different classical methods 4:4–8

Randive K (2013) Elements of geochemistry, geochemical exploration and medical geology

Rehan I, Khan MZ, Rehan K, Sultana S, Rehman MU, Muhammad R, Ikram M, Anwar H (2019) Quantitative analysis of Fuller's earth using laser-induced breakdown spectroscopy and inductively coupled plasma/optical emission spectroscopy. Appl Opt 58(16):4227–4233. https://doi.org/10.1364/AO.58.004225. PMID: 31251224

Satpute AD (2003) Rasaratna Samuchchaya (Translated in English). Chaukhamba Sanskrit Pratishthan, New Delhi, ISBN: 8170842062; 316

Savrikar SS, Ravishankar B. Introduction to 'Rasashaastra' the iatrochemistry of Ayurveda. Afr J Tradit Complement Altern Med 8(5 Suppl):66–82. https://doi.org/10.4314/ajtcam.v8i5S.1. Epub 2011 Jul 3. PMID: 22754059; PMCID: PMC3252715

Shahriari M, Zare F, Nimrouzi M (2018) The curative role of bitumen in traditional Persian medicine. Acta Med Hist Adriat 16(2):283–292. https://doi.org/10.31952/amha.16.2.6. PMID: 30488706

Sharma S (1998) New Delhi, India: Motilal Banarasida. 'Rasa Tarangini' 15/69–71

Sharma S (1998) New Delhi, India: Motilal Banarasidas. 'Rasa Tarangini' 15/81

Sharma S (1998) New Delhi, India: Motilal Banarasidas. 'Rasa Tarangini' 17/46

Sharma S (1998) New Delhi, India: Motilal Banarasidas. 'Rasa Tarangini' 18/39–42

Sharma V (2018) Vimal (Iron Pyrite): a medicinal mineral drug of Ayurveda—an approach to develop its mineralogical monograph. Asian J Pharm (AJP) 12(04) https://doi.org/10.22377/ajp.v12i04.2837

Shubha HS, Hiremath RS (2010) Preparation and physicochemical analysis of Rasaka Bhasma. AYU 31(4):509–510. https://doi.org/10.4103/0974-8520.82025. PMID: 22048549; PMCID: PMC3202248

Shukla A, Singh J (2019) Hingulottha parada: a traditional mercury extraction method. Int J Green Pharm 10:121–124

Singh, Mishra R (2019) Classical review of Gandhaka (sulphur)—an Ayurvedic perspective. Int J Unani Integr Med 3. https://doi.org/10.33545/2616454X.2019.v3.i3a.96

Singh D, Kailoria P, Sharma R (2022) A critical review on Hemagarbha Pottali: an effective Ayurveda formulation. AYUSHDHARA 50–54. https://doi.org/10.47070/ayushdhara.v9iSuppl2.1046

Somadeva (2004) Varanasi, India: Chaukhambha Orientalia. 'Rasendra Chudamani' 14/23

Sud S, Bandari S (2014) A systematic overview on—Parpati kalpanas 2:14–23

Sud S, Khyati S, Sud, Vinay T (2016) Underpinning the classical preparation Rasakarpura (mercurial salt)—a review. World J Pharm Pharm Sci 5:251–279. https://doi.org/10.20959/wjpps201610-7766

Sumersingh Rajput D, Gokarn R, Jagtap CY, Galib R, Patgiri BJ, Prajapati PK (2016) Critical review of *Rasaratna Samuccaya*: a comprehensive treatise of Indian Alchemy. Anc Sci Life 36(1):12–18. https://doi.org/10.4103/0257-7941.195412. PMID: 28182027; PMCID: PMC5255965

Upadhyaya M (1999) Varanasi, India: Choukhambha Bharati Academy. 'Ayurveda Prakaasha' 3/284

Vagbhatta R (1998) New Delhi, India: Meharchand Lachhmandas Publications. 'Rasaratnasamuccaya' 5/208

Wele A, De S, Dalvi M, Devi N, Pandit V (2021) Nanoparticles of biotite mica as KrishnaVajraAbhraka Bhasma: synthesis and characterization. J Ayurveda Integr Med 12(2):269–282. https://doi.org/10.1016/j.jaim.2020.09.004. Epub 2021 Jan 2. PMID: 33402266; PMCID: PMC8185975

Wijenayake A, Pitawala A, Bandara R, Abayasekara C (2014) The role of herbometallic preparations in traditional medicine--a review on mica drug processing and pharmaceutical applications. J Ethnopharmacol 155(2):1001–1010. https://doi.org/10.1016/j.jep.2014.06.051. Epub 2014 Jun 30. PMID: 24993885

Young S, Miller J (2019) Medicine beneath your feet: a biocultural examination of the risks and benefits of geophagy. Clays Clay Miner 65. https://doi.org/10.1007/s42860-018-0004-6

Zhao M, Li Y, Wang Z (2022) Mercury, and mercury-containing preparations: history of use, clinical applications, pharmacology, toxicology, and pharmacokinetics in traditional chinese medicine. Front Pharmacol 13:807805. https://doi.org/10.3389/fphar.2022.807805. PMID: 35308204; PMCID: PMC8924441